CW00385876

Hydropower Developments – New Projects and Rehabilitation

Organizing Committee

George Aggidis
Gilbert Gilkes & Gordon Limited
Kendal, Cumbria, UK

James Bain
Hydropower Engineering Limited
Orpington, Kent, UK

Angus Grant
Consultant
Doune, Perthshire, UK

IMechE
Seminar Publication

I MECH E

Hydropower Developments – New Projects and Rehabilitation

Held on 30 November 2000
At IMechE Headquarters, London, UK

Organized by
The Fluid Machinery Committee of the Power Industries Division of
The Institution of Mechanical Engineers (IMechE)

Co-sponsored by
The International Journal of Hydropower and Dams
British Hydropower Association

IMechE Seminar Publication 2000–17

**Professional
Engineering
Publishing**

Published by Professional Engineering Publishing Limited for The Institution of
Mechanical Engineers, Bury St Edmunds and London, UK.

First Published 2000

This publication is copyright under the Berne Convention and the International Copyright Convention. All rights reserved. Apart from any fair dealing for the purpose of private study, research, criticism or review, as permitted under the Copyright, Designs and Patents Act, 1988, no part may be reproduced, stored in a retrieval system, or transmitted in any form or by any means, electronic, electrical, chemical, mechanical, photocopying, recording or otherwise, without the prior permission of the copyright owners. *Unlicensed multiple copying of the contents of this publication is illegal.* Inquiries should be addressed to: The Publishing Editor, Professional Engineering Publishing Limited, Northgate Avenue, Bury St. Edmunds, Suffolk, IP32 6BW, UK. Fax: +44 (0)1284 705271.

© 2000 The Institution of Mechanical Engineers, unless otherwise stated.

ISSN 1357–9193
ISBN 1 86058 317 2

A CIP catalogue record for this book is available from the British Library.

Printed by The Cromwell Press, Trowbridge, Wiltshire, UK.

The Publishers are not responsible for any statement made in this publication. Data, discussion, and conclusions developed by authors are for information only and are not intended for use without independent substantiating investigation on the part of potential users. Opinions expressed are those of the Author and are not necessarily those of the Institution of Mechanical Engineers or its Publishers.

Related Titles of Interest

Title	Editor/Author	ISBN
IMechE Engineers' Data Book – Second Edition	C Matthews	1 86058 248 6
Handbook of Mechanical Works Inspection	C Matthews	1 86058 047 5
Wind Energy 1999 – Wind Power Comes of Age	P Hinson	1 86058 206 0
Power Station Maintenance	IMechE Conference	1 86058 274 5
CHP 2000: Co-Generation for the 21st Century	IMechE Conference	1 86058 141 2
CCGT Plant Components – Development and Reliability	IMechE Seminar	1 86058 190 0
Hydro Power Developments Current Projects, Rehabilitation, and Power Recovery	IMechE Seminar	1 86058 121 8
Plant Monitoring and Maintenance Routines	IMechE Seminar	1 86058 087 4

For the full range of titles published by Professional Engineering Publishing contact:

Sales Department
Professional Engineering Publishing Limited
Northgate Avenue
Bury St Edmunds
Suffolk
IP32 6BW
UK

Tel: +44 (0)1284 724384
Fax: +44 (0)1284 718692
Website: www.pepublishing.com

Contents

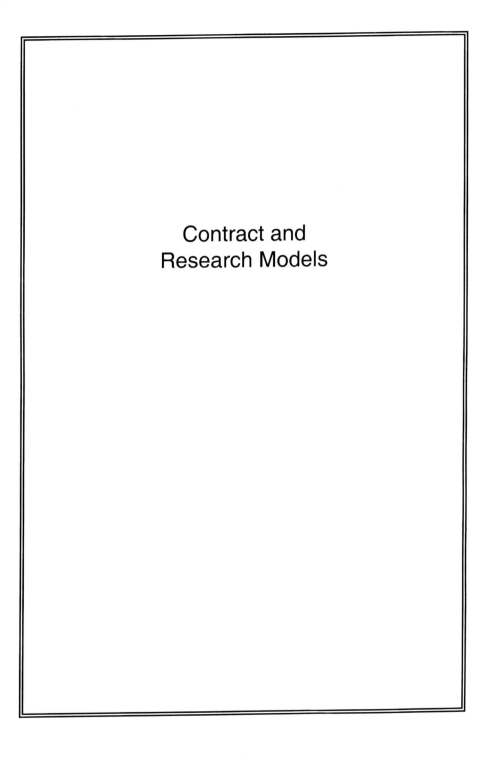

Contract and
Research Models

S724/002/2000

Detailed pressure survey of a lattice valve seal under high-flow conditions

G L LONG and **P T IRELAND**
Department of Engineering Science, University of Oxford, UK
M J RIGBY
GE Energy (UK) Limited, Doncaster, UK

SYNOPSIS

Large butterfly valves have been supplied for hydropower stations by GE Energy (UK), formerly Kvaerner Boving Ltd, since 1931. For a description of a recent project that includes the application of large butterfly valves see Jones and Taylor (2000). Some of the most demanding operating conditions for a power station protection valve may be experienced during emergency closing. The high flow rates and large dynamic heads during such conditions can cause large variations in the static pressure around the valve disc and the resulting loads on valve components can be immense. The prediction of these loads presents the valve designer with a challenging problem, even with modern tools such as finite element analysis and Computational Fluid Dynamics. This paper reports the development of a means of predicting the loads whilst closing on a valve disc rubber seal in a high head power station protection valve. GE Energy performed the research in collaboration with the University of Oxford to establish the fundamental design parameters and to evaluate seal loads and deflections. The research reported here focused on quantifying the pressure loads on the seal section of a lattice valve during high flow closing. Detailed static pressure measurements were made on a scale model of the valve over the location of the seal. Dimensional analysis allows the real system loads to be predicted from the model tests. The results have shown that the complex separated flow while closing develops unexpectedly high pressure forces on the seal.

1. INTRODUCTION

Lattice blade butterfly valves may be employed in hydropower stations, for example, as turbine inlet valves or protection valves such as penstock protection valves. In many hydroelectric power stations it is necessary to include a means by which water can by-pass the turbine stage, when power is not required, in order to maintain a flow downstream of the power station. To achieve this a by-pass pipe line is normally provided which terminates in an energy dissipating valve. A lattice blade butterfly valve may be installed a short distance upstream of the energy dissipating valve. The valves have two functions. The first is to allow maintenance of the energy dissipating valve or turbines without having to de-water the upstream pipeline. The second is to act as an emergency shut off valve in the case the energy dissipating valve or turbine fails thus avoiding emptying the reservoir and flooding of the downstream area.

When the butterfly valve is closed for maintenance the hydrodynamic loads are low since the energy dissipating valve has already been closed. However, when the butterfly valve is operated in the emergency condition, it is closed against high flow water rates. Under these conditions, the rubber seal on the downstream disc of the valve experiences high loads and there have been cases where the seal extrudes from its housing and subsequently prevents the valve from closing fully.

2 SEAL TECHNOLOGY IMPROVEMENTS

The development work investigated both the suitability of alternative rubber seal ring designs and the prediction of seal loads during flow.

2.1 Rubber seal alternative designs
The use of rubbers with different stiffnesses and fibre reinforcement was investigated using finite element analysis, (Partridge 1998). Several seal configurations were analysed with the aim of determining a seal design capable of resisting the hydrodynamic loads during closing but not requiring excessive clamp-ring bolt loads to partially extrude the seal during fitting on site. During this analysis, it became clear that the pressure boundary conditions acting on the seal surface during closure were unknown. It was decided to research the pressure loads with the aim of enabling the seal design to be improved

2.2 Seal load measurements
A scaled down (approximately 1/8th full size) model of the valve was used to obtain seal loading data during closure. The rig was designed, built and tested at the University of Oxford. Unnecessary details were not reproduced but, on the whole the model is a faithful small-scale reproduction of the actual valve. In particular, the exact form of the seal was machined into the aluminium disc representing the downstream disc fitted with the rubber seal. The results obtained give a high-resolution description of the pressure forces acting on the seal during valve operation in high flow conditions. Dimensional analysis allowed the results to be translated to any valve with the same geometry.

3 EXPERIMENTS

3.1 Dimensional analysis
In the absence of cavitation, dimensional analysis shows that the flow field is a function only of Reynolds number. This simplification is acceptable since there is no cavitation associated with the full size valve over its full operating envelope. The stresses from the fluid are transmitted to the surface of the seal as pressure (normal to the seal surface) and shear stress (tangential to the surface). Order of magnitude analysis using standard correlations indicated that the shear stress acting on the rubber would not be sufficient to cause the seal to deflect

significantly. The more important load is the pressure load. Furthermore, physical insight suggested that the seal shape depends predominately on the pressure variation around the seal and not the absolute pressure. . When the valve is operated in the emergency condition it must close against a flow which is initially $32.5 m^3 s^{-1}$. The latter gives a pipe Reynolds number in excess of 20million. The further simplification is made that the core flow (outside viscous effects and boundary layers on the valve component surface) is insensitive to Reynolds number. This means that scale model tests can be performed at lower than power station Reynolds numbers. In other word, any measured pressure coefficient (presented below) is the same for the model as for the real valve. In fact, to improve the accuracy, the measured pressure coefficients use the pressure difference across the valve as the denominator. Note the above shows that it is acceptable to change the working fluid to air provided (i) the model flow Mach number is low so the flow is incompressible (ii) the air flow is still turbulent (iii) cavitation effects are not present or significant in the real application.

The pressure drop across the full size valve was predicted using a Flow-Master analysis (King, 1998) that modelled a complete installation at Lesotho, (Jones and Taylor 2000). In the latter prediction, the valve loss coefficient (as a function of valve closure angle) had been determined by separate model tests. Good agreement was confirmed between the loss coefficient measured on the current air flow model and the results from the earlier tests. The measured model pressure coefficients could then be used to predict the full size valve pressures during closing.

3.2 Experimental Apparatus

Figure 1 Model lattice valve showing the perspex housing construction.

The valve body was machined from a solid block of perspex sectioned along the centreline into two halves to allow easy access to the valve discs. Figure 1 shows the model valve both fitted to the tunnel ductwork and also with the upper half removed for access to the discs. Sufficient pressure tappings were included to enable the static pressure over the downstream seal surface to be resolved at high resolution. Figure 2 shows a detail of the actual rubber seal and clamp ring arrangement. Note that the real seal components include the valve disc, the rubber seal and the clamp ring. The model downstream, pressure tapped disc was designed to produce an accurate small scale duplicate of the assembly surface geometry. To improve the experimental accuracy, for each pressure reading, the pressure difference between the seal tapping and the pressure at the location of the seal ring to disc joint with the downstream face of the disc was measured directly. For this reason, for each tapping position indicated in Figure 3, a separate tapping at the location marked J in Figure 2 was included. The model valve discs, valve ribs and trunnions were manufactured from aluminium.

The upstream disc geometry was machined with simplified geometry. The downstream disc periphery, where the rubber seal is located in the real valve, was modelled with a profile that was an exact scaled representation of the full size disc. Pressure tappings were place at $10°$ intervals across the top and bottom and of the disc and at different depths or distance from the flat face as shown in Figure 3. The downstream disc was designed so that an outer ring which include the pressure tappings could be rotated about the axis of the downstream disc. This allowed the static pressure to be determined over several rings around the disk to produce a full map of disc surface pressure.

The pressure tappings were included at 5 disc depths to ensure that a detailed map of the disc periphery pressure could be obtained. Tapping 6 is at a position corresponding to the rubber seal surface and some data from this position are presented and discussed below. Two angular rotations need to be given to specify the valve state and the tapping location.

Valve position	Zero when valve fully open, $90°$ when valve closed.
Tapping rotation	Zero when tapping at highest location for valve top tappings. Zero when tapping at lowest location for valve bottom tappings $90°$ when tappings are adjacent to trunnions.

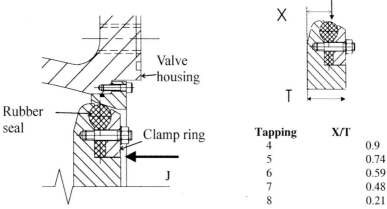

Tapping	X/T
4	0.9
5	0.74
6	0.59
7	0.48
8	0.21

Figure 2 Detail of the rubber seal arrangement in the real valve.

Figure 3 Tapping axial position

The tapping pairs were connected to the pressure transducer, via a switchbox, using 1mm outer diameter polythene tubing that followed internal routes through the valve disc and valve trunnions. A centrifugal fan drew from atmosphere through the valve. The upstream and downstream pipe lengths were chosen to match standard GE Energy model tests.

4. RESULTS

4.1 Data processing

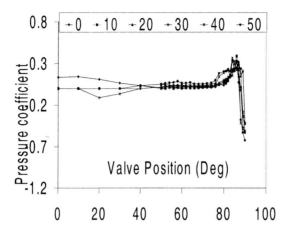

Figure 4 Pressure coefficient for tapping 6 as a function of valve closure angle at different tapping rotations from top central position. A positive coefficient corresponds to push out force

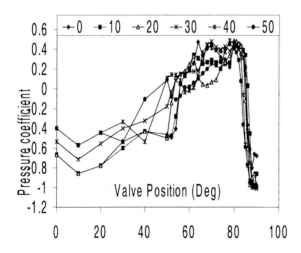

Figure 5 Pressure coefficient for tapping 6 at different rotations from bottom centreline position. A positive coefficient corresponds to push out force

Example plots of the dimensionless pressure coefficients for one of the tappings at the rubber seal location are shown in Figure 4 and Figure 5. Note that the pressure coefficient is the difference between the pressure downstream at the seal clamp ring joint and the local tapping pressure divided by the pressure difference across the valve. By multiplying each of the dimensionless coefficients by the corresponding pressure drop across the full scale valve from the Flow Master prediction, Figure 6, the real seal loads were calculated. The resulting plots of the pressure difference acting on the top of the seal are shown in Figure 7.

The results for the top of the valve, Figure 4, show that the push out force acts on the seal over a valve closure range from about 80° to 87°. A 2 dimensional I-DEAS animation showing the change in geometry of the space between the valve top and the casing was generated and examined in detail,. The animation showed that, when the valve is almost closed, there is a gradual increase in the area at the top of the valve as the flow passes over the valve disc. The pressure difference between the seal surface and downstream is due to flow diffusion through the gap between the valve disc and valve body, giving rise to an increase in static pressure downstream. Figure 4 also shows that the push out force begins to act at the sides of the valve (with the tapping rotated to 40° and50°) before the top (0°) This is to be expected as the sides of the valve disc come into close proximity with the valve body before the top. When the valve reaches 87° the push out force reverses, as the seal gets very close to the valve body. The increased pressure on the top of the seal is most likely due to a viscous pressure drop between the very small gap.

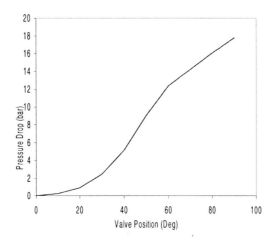

Figure 6 Predicted pressure difference across the real valve

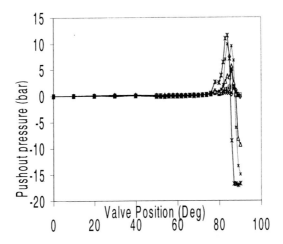

Figure 7 Predicted real system pressure difference at top centreline rotation for different tappings

The results for the bottom of the valve, Figure 5, show that up to a closure of 55° there is a push in force acting on the seal. This is due to the flow dividing streamline being on the disc surface near the tappings. After 55° there is a sharp transition to a pull out force acting on the seal at the bottom of the valve. This indicates that the dividing streamline has changed position and moved up the seal surface so the flow must make a sharp turn before passing under the disc. As the valve comes to the fully closed position the force acting on the seal at the bottom of the valve changes back to a push in force. This is most likely due to viscous forces producing a pressure difference in the small gap.

5. PUSH OUT DEMONSTRATION

The magnitudes of the pressure differences acting across the seal due to the diffusion of the flow are of the order of 13bar. Separate hydrostatic tests were performed on a rig designed (Beland, 1998) to represent a section of the seal and valve to investigate the effect of water pressure on the clamp ring joint. The tests showed that, for standard clamp ring nut torques, a pressure difference of 12.5 bar was sufficient to push the rubber seal from its seat.

6 CONCLUSIONS

A small scale, pressure tapped model of a lattice valve has proved very useful in an study of seal loads. Experiments showed that significant forces act on the seal due to the interaction of the valve disc geometry with the valve body geometry as the valve closes. The test strategy is being used by GE Energy as part of an ongoing valve research and development programme.

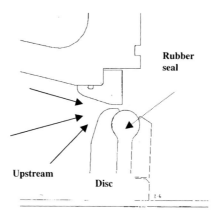

Figure 8 Detail of the aperture geometry as the seal closes. Valve is at 2.6°.

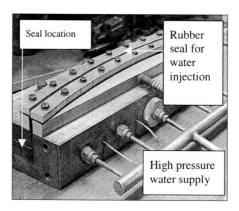

Figure 9 Rig for hydrostatic tests.

7. ACKNOWLEDGEMENTS

The authors gratefully acknowledge GE Energy for supporting this project. The model valve was manufactured by Mr. Geoff Horton at the Dept. Engineering Science and his technical expertise was much appreciated.

8. REFERENCES

Beland, R., 1998," Investigation into methods to resolve by-pass guard valve seal being extruded under emergency closure," Kvaerner Boving Report 26518.
Jones, G. C and **Taylor, R.G.,** 2000, "Design and supply of mechanical equipment for the Muela power station, Lesotho," submitted to I.Mech.E conference "Hydropower Developments – Current Projects, Rehabilitation and Power Recovery" S274
King D, 1998, "Muela bypass information" KBO document 23385.
Partridge, I., 1998,"Muela hydropower project" Kvaerner Boving Report 27063.

© 2000 GE Energy (UK) Limited

Component Development

S724/003/2000

Fit for pressure – an investigation into the behaviour of 'Piano Note' elastomeric seals

H M K DICKSON
Weir Pumps Limited, Glasgow, UK
M C JONES
Edison Mission Energy, First Hydro Company, Llanberis, UK

1. INTRODUCTION

Within Hydro Electric Plants, a variety of gate designs are employed to control water levels in dams or provide isolation, either to shut down turbines in an emergency, or permit de-watering so that inspection/maintenance work can be carried out safely. The clearance between the moving Gate and, either the Dam Abutments, or the Portal Frames, has to be sealed to stop leakage. If the seal does not work effectively the water escaping can have consequences which may vary in effect from inconvenience to critical. In the case of Drum Gates, the leakage can eventually damage the structure. However in the case of Isolation Gates, the leakage is a potential hazard to engineers working downstream of the gate in the de-watered tunnel(s).

The range of pressures over which the seals must operate can vary from zero, at the surface of the water, to some 200 feet below the free surface. The distance over which the gates must be lowered to the portal to effect closure also varies by up to about 200 feet. The seals have therefore to work efficiently from low to high differential heads and traverse long lengths of portal abutment.

1.1 Historical Perspective
Over the years several types of seal have been used in such locations to cope with the range of conditions. The types employed include rolled brass strip Figure 1, "L" section fabric reinforced elastomer Figure 2, rectangular section rubber Figure 3, and more recently "Piano Note Section" extruded elastomer Figure 4. The way these seals have been mounted has traditionally been based on hard won experience and varies from application to application.

1.2 Typical Applications
If we examine the extreme installations, Low Head surface gates and High Head tunnel gates in detail, we can see how the varying operating parameters affect the behaviour of the seal.

1.2.1 Low Head Surface Installations

In the case of Drum Gate Seals, Figure 5, the differential head is low and the gate has to move regularly to control the level of the impounded water behind the dam. The seal has thus to work with a very low hydraulic load, seal effectively, and withstand relative motion over the side abutment face of the dam structure. In such applications some pre-load is required to ensure an efficient seal and the abutment is normally faced with a stainless steel sheet to provide a smooth and durable surface over which the seal slides during operation of the gate.

1.2.2 High Head Tunnel Installations

In the case of a Deep Tunnel Isolation Gate seal, Figure 6, the differential pressure across the closed gate is high, but generally operation of the gate is infrequent. Normally this is when tunnel inspections are required but the gate may also be shut automatically in an emergency. The seal has therefore to work under a higher hydraulic head, seal very efficiently and withstand motion over the face of the portal frame as the gate is lowered and raised. Because some of these gates can be up to 10 metres high, if the seal is installed with a pre-load, the seal contact line of the seal is scrubbed up and down the portal face resulting in either wear or tearing of the seal. In deep installations, when the gate is closed, the differential pressure across it can be high enough to press the seal firmly against the portal face so that it is not necessary to have any initial pre-load to ensure a tight closure.

Where the seals have been installed with pre-load it is quite common to find damage to the sealing face when carrying out surveys prior to overhauling the gates. This wear can be quite significant with abrasion down the length of the contact line on the seal.

1.2.3 The Operating Depth at which Seal Pre-Set is not required

There is an operating depth beyond which the differential pressure across the seal is sufficient to deflect the bulb onto the portal and there is no need to have any initial pre-setting load. For gates operating below this level, the seal can be installed without any pre-set, i.e. it can be installed back from the plane of the portal frame. If this is done there will be no contact between the seal and the portal face until the gate has almost been lowered to its working position. At this point the gate restricts the water flow sufficiently for the differential pressure to build up and "flip" the seal into contact with the portal frame and stop any flow.

2. DESCRIPTION OF THE OPERATION OF PAST AND PRESENT SEAL TYPES.

2.1 Rolled Brass Strip, Figure 1

Thin Brass strip rolled along one edge to form a "Tensator" like coil has been used as a seal. The rolled edge was resilient to flexing and the brass formed a reasonably good rubbing combination with the portal frame material. The low Inertia of the brass section meant that it deflected readily under differential pressure to effect an efficient seal. However, the thinness of the brass strip which gave it its resilience meant that it was susceptible to degradation by wear.

The effectiveness of the seal depended firstly, on how well the seal was fitted, secondly on the condition of the portal face, and finally on how well the seal displaced the deposit of peat which can build on the portal face over time. In addition to this, cases of differential corrosion have been noted with brass fixing screws causing the seal to fail.

2.2 "L" Section Fabric Reinforced Elastomer, Figure 2

This has generally replaced the brass strip but it is not nearly so flexible. Since the section is thick and is fabric reinforced it can withstand significant wear but the lack of flexibility means that it is not as responsive to changes as the seal wears, Photograph 1.

The stiffness of the seal section often meant that setting the seal in the optimum position was achieved by trial and error. Generally this seal is not efficient at wiping away peat deposits from the portal faces as the gate is lowered into position and this affects the efficiency of the "seal".

2.3 Rectangular Section Rubber, Figure 3

This seal is normally used along the bottom of gates. The simplicity of installation and operation is only compromised by a lack of resilience if there should be any lack of flatness of the mating sill or there is debris on the sill.

2.4 "Piano Note Section" Extruded Elastomer, Figure 4

This seal was introduced because it was a robust section, which also incorporated a degree of self-adjustment because of the inherent flexibility of its section profile and pure elastomeric composition.

Because the seal is made from unreinforced extruded elastomer it can conform well to deviations in the portal face and provide a tight seal. Because there is "line contact" between the bulb of the seal and the portal face any build up of peat on the portal face tends to be pressed out from between the contacting surfaces. While the Piano note seal has performed well in service, when documented performance information was sought, none of the suppliers approached could provide any test data. It was then clear that independent tests would have to be carried out to determine the operating parameters of this type of seal and to explore its behaviour under various conditions.

The seal section can be made more resistant to abrasion if PTFE inserts are fitted down the contact or sealing line, Figure 7. However while this makes the sealing point more resistant to abrasion it stiffens the section so that it does not conform so readily to the portal face and as a result the "seal" is not so efficient. In addition to this the keying of the PTFE into the seal profile has been known to fail in service. On balance it is better to use the pure elastomeric section.

3.0 REQUIREMENTS FOR NEW OR REPLACEMENT SEALS

Naturally the Power Utilities who operate Plants want a new seal that will be longer lasting and provide an equivalent or superior performance to the seal being replaced. In addition to this, they now seek some form of guarantee or documented assurance that the new seal will indeed be better than the old one.

3.1 The Incentive for Conducting Seal Tests

Before a supplier can offer such assurances on the operability and reliability of a seal, he must understand both the behaviour of the seal or seals that are being offered, and the conditions under which they must operate. A market survey showed that no load deflection characteristic testing of seals had been performed by any of the suppliers approached. Accordingly a programme of tests had to be set out which would enable us to explore the behaviour of the seal section, both along the sides of the gate, and at the corner joints. The first tests looked at a sample representing a section of seal from either the side or top of the gate and the second tests looked at the behaviour at a corner where the seal sections along the side and the top of the gate meet.

Only by conducting such tests, was it possible to explore the behaviour of the seal and then predict its behaviour at the various differential water pressures on site. The data obtained from the tests was then used to support the assurances given to the Power Utility, on the operability and reliability of the proposed new seal installation. It could also be used to prepare procedures for the correct fitting/installation of the seals and also subsequent inspections of the seals at a later date.

3.2 Defining the Parameters to be Investigated in the Tests
The agreed test programme allowed a number of parameters to be explored, and these are addressed in detail as follows.

3.2.1 Simulation of a Suitable Range of Operating Differential Pressures
A simple test rig using a "Tensile Testing" machine was set up to conduct the tests, Photograph 2. This was capable of applying a wide range of loads to the test samples of seal and so to mimic the full range of pressures encountered at Power Plants. The loading could be applied smoothly and accurately by the machine, which also gave an automatic display of the seal deflection. Because the rig was open, the general pattern of the deflection of the seal could be observed at all times and photographed as each test progressed.

3.2.2 The Necessity for Inserting "Seal Clamping Stops" during Installation of the Seals
Traditionally, Bronze or Hard Nylon spacers have been fitted around the clamping bolts where these passed through the flange part of the Piano Note seal. The reason for this was to stop over compression of the flange of the seal section and thus avoid either damage to the rubber or prevent over compression of the flange section that might inhibit correct functioning of the seal.

3.2.3 The Effect of Different Levels of Backing Support Behind the Seal on the Seal Flexibility
It was anticipated that when the degree of support behind the flange section of the seal was reduced, the "Section" would deform more readily, i.e. require a smaller force to deflect it, so that lower differential pressures could deflect the bulb of the seal into contact with the portal face.

3.2.4 The Stiffening Effect on the Seal by the Formation of a Single Mitred Bonded Corner Joint
To ensure the most efficient seal around the periphery of a gate, the sections of the seal along the side and top of the gate are normally joined by the insertion of a factory made pre-formed bend. These can be bulky and mechanically stiff. An alternative design to this, a compact mitred joint, which could be formed "on site", was included in the test programme.

3.2.5 The Investigation of Modifications to the Simple Mitred Joint and an Alternative Compound Mitred Joint

Both the factory pre-formed corner piece, and the simple mitre joints, are by their construction stiffer than the base strip seal. Because of this it was important to investigate the degree of stiffening attributable to these corners and see if it was possible to minimise this so that the applied force required to effect a water tight seal could be brought as low as practicable.

4. TESTING AND RESULTS

4.1 Test Rig

The straight test piece of seal was loaded through simple wood plates, Figure 8, since this was simpler to operate and easier to control than a rig using water pressure. The sample corner test piece was also loaded by simple wood plates but the plates for each leg of the corner were in two sections so that load could be applied, not just over the full length of each leg, but both the inner and outer halves of each leg could be loaded separately. This provision allowed readings to be taken which could later be studied and enable the stiffening effect of the corner joint to be assessed.

All the sample test sections were mounted against moveable backing plates, Figure 8, which allowed the effect of the degree of support given to the seal on the seal's stiffness to be investigated.

4.2 Results

The first feature to be noted from the tests was the imperceptible effect that the degree of tightening of the clamping bolts had on the flexibility and hysteresis of the seal in operation, Figure 9.

The second feature to be noted was that, as expected, when the degree of support behind the flange of the seal was reduced, the seal became easier to deflect, Figure 10.

The third feature to be noted was the very significant coupling between the sections of seal when they were bonded together at a corner joint, resulting in exaggerated deflection of the "nib" of the corner, Figure 11.

5. DISCUSSION OF THE RESULTS

5.1 The Effect of the Magnitude of the Clamping Force on the Seal Flange on the Behaviour of the Seal

As stated above, 3.2.2, it has been common practice to fit collars or spacers around the clamping bolts where they pass through the rubber flange of seals. The intention behind this practise was firstly to prevent over compression of the rubber and secondly by limiting this, avoid over compression of the flange from affecting the performance of the seal.

In tests carried out, in connection with other investigations (1), sample sheets of rubber of varying thicknesses were compressed to measure their compressive stiffness. It was found that if plain sheets of rubber were compressed the stiffness rose rapidly as the load was applied and the material then behaved as a solid body with negligible elasticity. Only when a pattern of holes was cut in the sheet, and there were spaces for the rubber to press into, was there any reasonable range of deflection under load.

It was anticipated therefore that the rubber flange under the seal clamping plate would only be expressed laterally, and at that only to a limited degree, so that the flange would rapidly become a bound block under isostatic compression. Thus in the tests when the characteristic of the seal was measured firstly with the clamping bolts only "finger tight", and then with the bolts tightened up as far as practicable with a spanner, little difference in the readings was expected. The results in Figure 9, show that there was virtually no change in the behaviour of the seal between the methods of clamping the seal confirming the initial expectations.

5.2 The Effect of the Extent of Support for the Seal Flange on the Flexibility of the Seal under Load
From Photograph 3 it can be seen that as the seal deflects, the profile pivots about the knuckle, and the material behind this arches up. This mode of deflection requires quite a degree of force to cause deformation since more than simple bending is involved.

If, however, the edge of the backing plate is retracted, Photograph 4, the part of the flange of the section next to the knuckle flexes by simple bending offering less resistance to deflection, in addition, the seal can rotate more freely and larger displacements are possible. By allowing the seal to deflect more freely the mechanically applied load or water differential pressure required to "flip" the seal is significantly reduced. This will allow the seal to become self-sealing at lower differential pressures thus widening the range of potential applications.

5.3 The Exaggerated Deflexion of the Nib of a Corner Junction and its Effect on Sealing Performance
As indicated above, 3.2.4, it is normal practice to fit either pre-formed corner pieces or to mitre the seal sections at the corners of a gate and to bond them so that there are no discontinuities around the periphery of the gate, thus achieving the tightest possible seal. However, the tests showed that when the seal deflects the bulb section rotates. When the lengths of seal, on either side of the corner are connected by bonding, the rotational deflection in each is transferred to the end of the other. That part of the seal adjacent to the corner, therefore, deflects more that the rest of the seal and comes into contact with the portal face prematurely, Photograph 5. To force the rest of the seal to come into contact with the portal face, additional load (pressure) has to be applied, either that, or the exaggerated deflection of the "nib" of the corner has to be countered by relieving the contact line local to the corner. Alternatively, another method of forming the corner junction had to be found so that the corner would deform by a similar amount to the rest of the seal.

5.4 Steps to Improve the Sealing Performance of the Corner Junction
There are several possible ways, in which the exaggerated deflection of the "nib" of the seal may be countered, two of these are;-
Proportional dressing of the sealing line adjacent to the junction.
Modifying the degree of deflection transfer and magnification that takes place at the junction.

5.4.1 Proportional Dressing of the Contact Line Adjacent to the Junction

This method is deflection specific. If initial gap from the seal to the portal face gap is less than has been anticipated when the proportional dressing was carried out, the corner will not come into contact and there will be leakage through the gap. Conversely, if not enough proportional dressing has been done, the corner will still make premature contact, and there will be a gap next to the junction and again there will be a leak path. For this reason Proportional Dressing is not considered to be a practical solution to the problem.

5.4.2 Modifying the Degree of "Deflection Transfer" that takes place at the Junction

There are many ways, in which this could be achieved including.

Using a more complex mitre joint, fitting one or more insert sections.

Fitting a supple sandwich piece in the junction.

In the tests that were carried out, an additional section of seal profile was inserted at the junction, Photograph 6 and Photograph 7. The effect of this change on both the magnitude of the closing force and the lag between first contact and full "sealing" contact is shown in Figure 10.

5.5 Possible Changes in Seal Geometry to Improve the Performance

5.5.1 Changes to the Basic Profile of the Seal

To achieve the improved flexibility with the currently available seal profiles either:-

A thin plate has to be fitted under the flange of the section to allow the section to deflect as described.

Or, A step has to be formed in the sides of the gate supporting the seal.

However, an alternative would be to incorporate the "pillow" in the seal section. This can be achieved by either incorporating a small step in the die used for extruding the seal, or a strip of rubber could be bonded onto the cured extrusion. It transpires that attempts have been made to incorporate the pillow in the extrusion, but these were not successful because it was not possible to support the section as it cured following extrusion. Since carrying out the work reported in this paper, we have now found that at least one seal supplier successfully bonds strips of rubber to the seal section after curing.

It has been done to achieve the benefits we have recorded, but without carrying out any tests such as we have described to determine the quantitative effects of the modification to the seal.

By incorporating the "pillow" in the section supplied to the end user, fitting the seal becomes much simpler and errors in positioning the "pillow" are eliminated.

5.5.2 The Possibility of Eliminating "Deflection Transfer" between the Two Parts of the Mitre Joint

It would be interesting if additional tests could be carried out to see how easy it might be to:-

Fit a supple link, a slice of softer rubber, into the mitre without affecting the performance of the seal.

Develop a new corner piece with a more supple profile.

5.6 Optimisation of the Fitting of the Seal to suit variations in Operating Parameters

The curves shown in Figure 11 show how by altering the extent of support behind the seal the flexibility is changed. This data can be used to calculate the differential pressure required to cause the seal to "flip" and seal against the portal face. It will allow engineers to determine the conditions under which the seal can be installed without pre-load and thus reduce the wear and tear on the seal during lifting and lowering gates. We believe this will significantly extend the life of seals for minimal investment.

5.7 The way forward

By the time this paper is published, seals with pillow plates will have been fitted on the Main and Stop Gates at the Dinorwig plant belonging to Edison Mission Energy. This will have been done as the result of the investigations described in this paper. It is expected that the seals will perform as predicted and that the damage previously experienced on the seals of these gates will have been effectively eliminated.

It is also expected that when the seals on other gates, main, tailgate and stop log, require replacing, this will be done in a similar manner thus offering improved seal life and operability.

It is envisaged that seal profiles incorporating a small pillow in the mounting flange, to simulate the support plate used in these tests, which are offered by some suppliers, will be used regularly on the gates for hydroelectric and other projects. As indicated above, this improved seal section will offer End Users more adaptable seals, which are just as easy to fit as the unmodified seal profile. Seals of this type will offer longer life and a greatly reduced risk of damage to the seal during operation of the gates.

While the work covered in this paper was initiated to validate proposed changes to the sealing on the gates at a particular hydroelectric plant, the philosophy should apply to other seal applications. These will include other types of gate associated with hydroelectric plants, canal lock and barrage gates, and also large flap valves (gates).

ACKNOWLEDGEMENT

The authors wish to thank the directors of their respective companies for permission to publish and present this paper.

REFERENCES

1. In-house investigations by Weir Pumps Limited into the use of elastomeric pads under the mounting feet of pumps to provide either vibration isolation and or de-tuning from natural frequencies of foundations.

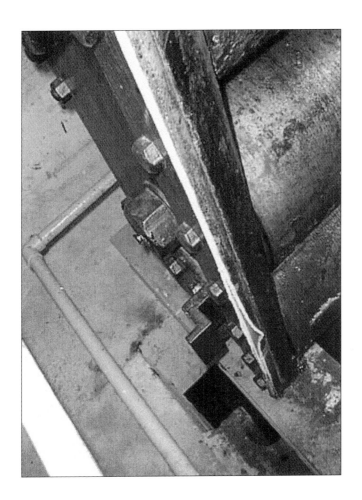

Photograph 1

Photograph 2

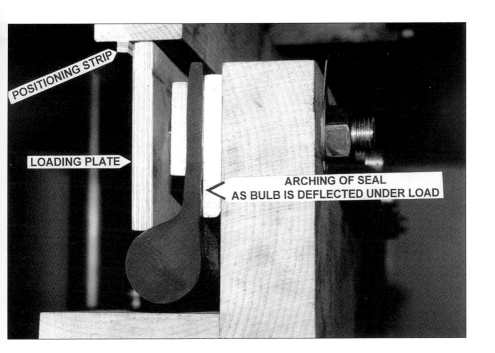

POSITIONING STRIP

LOADING PLATE

ARCHING OF SEAL
AS BULB IS DEFLECTED UNDER LOAD

Photograph 3

Photograph 4

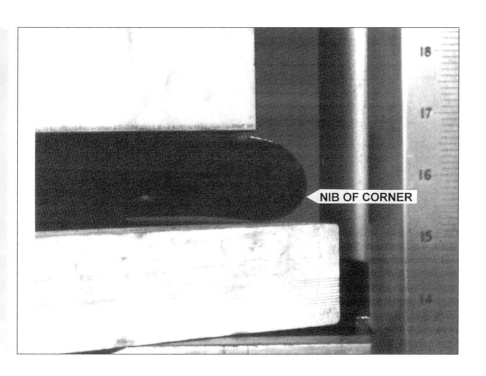

Photograph 5

Photograph 6

S724/003/2000 © IMechE 2000

Photograph 7

S724/004/2000

High-pressure components review at Ffestiniog pump storage power station

I PARTRIDGE
GE Energy (UK) Limited, Doncaster, UK
D P MORGAN
Edison Mission Energy, First Hydro Company, Llanberis, UK

SYNOPSIS

The Ffestiniog Pump Storage Power Station is situated near Blaenau Ffestiniog in North Wales, U.K. Commissioned in 1965, the station continues to operate in turbine and pump mode, resulting in frequent pressurisations and depressurisations. Since the station started operation the units have undergone approximately 90,000 turbine pressurisations, which is considered to be a typical design. In 1996 First Hydro as new owners commissioned G.E. Energy (UK) Ltd (formally Kvaerner Boving) to undertake a study into the continued life of the station.

The paper discusses work carried out to review the major hydraulic pressure components with an operating target of another 75,000 turbine pressurisations thus extending the station life by a further 25 years. The study had to take into account the material findings from a catastrophic failure, which had occurred during pressure testing of one of the turbine spiral casings in 1961.

The paper discusses, defects which were propagating to a level of concern on two of the units, in the highly stressed manhole area of the turbine spiral casing. The resulting repair of these defects had to take into account the 1961 casing failure which had initiated from an area which had been weld repaired.

The paper will help operating engineers who are contemplating refurbishment and life extension beyond the design life. It is a case study to show the many considerations that have to be evaluated in order that correct engineering decisions are made.

INTRODUCTION

First Hydro operate Dinorwig and Ffestiniog pump storage power stations, which are located 25 miles from each other in Gwynedd, North Wales, UK.

GE Energy (UK) Ltd. Based in Doncaster, UK., have many years experience of the design and installation of hydropower plant world-wide and were the original equipment manufacturers of the Dinorwig pump turbines.

FFESTINIOG POWER STATION

➢ Four Francis turbines and four twin suction pumps rated at 75 MW each
➢ First pumped storage power station in the UK at Blaenau Ffestiniog, North Wales.
➢ Commissioned in 1965.
➢ Turbine later re-rated to 90 MW each.

See figure 1 for unit section

BACKGROUND HISTORY

During the construction of Dinorwig Power Station, also operated by FHC, the high pressure components were individually subjected to engineering critical assessment, (ECA) so as to determine the non destructive examination NDE inspection frequency. Extensive non-destructive examination was carried out during manufacture and site installation. Having the information, from this assessment, gives the operator the confidence to continue with the long term operation of the high pressure components.

At Ffestiniog, the historical information relating to the design and manufacture of the high pressure components no longer existed. It therefore became apparent that a programme of extensive testing, followed by an ECA to establish NDE inspecting frequency, was required so as to give continued safe operation in the future. This paper defines the basis of the work carried out as part First Hydro's asset management programme.

The turbine spiral casings manufactured for Ffestiniog were subjected to hydraulic test at twice the working pressure. The spiral casing destined for Unit No. 3 turbine, failed catastrophically during works pressure test (see figure 2). The application of present day fracture mechanics theory, would give an understanding as to why the original failure had occurred and would provide valuable information for future safe operation of the plant.

REASONS FOR THE REVIEW AND SCOPE

➢ Low cyclic fatigue due to continued pressurisation and depressurisation of high pressure components.
➢ Original design life expectancy in the order of 30 years.
➢ Approximately 90,000 turbine pressurisation carried by 1996

- Continued safe and reliable operation of Ffestiniog for a further 25 years is dependant on the integrity of the major high pressure components especially embedded parts.
- Scope: Turbine spiral casings, pump spiral casings, main inlet and Pump discharge valves (See figure 3).

COMPONENT REVIEW OBJECTIVES
- The primary objective of the review of the high pressure components is to demonstrate, that the plant can be safely operated for the next 25 years
- 25 years future operation equates to approximately 75,000 turbine pressurisations
- At the end of the 25 years the components should still be able to withstand the worst case over pressurisation ~ equivalent to a 90 MW trip

PHILOSOPHY AND STAGES FOR REVIEW OF HIGH PRESSURE COMPONENTS

Accurately quantifying the remaining life and condition of a component or plant can only be achieved by the completion of an Engineering Critical Assessment. The ECA analyses the components / plant, to evaluate its future fitness for purpose for its given predicted operation and environment. This evaluation is carried out by examining the structure with respect to; plastic collapse, fracture mechanics, fatigue, corrosion, erosion and wear.

To facilitate a ECA the review was broken down into three key stages;

Stage 1   Collection of known data ~ original drawings, historical loading, future loading, modifications etc. The research and collection of the above information can prove challenging particularly historical data, which may not be available or archived.

Stage 2   Non-destructive examination (NDE) of the components by Ultrasonic (UT) and Magnetic Particle (MPI) techniques. Thickness and geometric checks were also carried out.

Stage 3   Examination of the components via an Engineering Critical Assessment (ECA). The integration of data from the first two stages along with detailed finite element analyses results allows assessments to be made of the components. The assessment is split into two distinct areas; Firstly the analysis of known defects and secondly the analysis of the remaining structure. Once this has been completed accurate and safe conclusions on components remaining life can be reached. If the component fails the assessment acceptance criteria then remedial action would be required.

REVIEW RESULTS TO DATE

As Ffestiniog is a working station the NDE has been carried out when dewatered outages were available. The corresponding schedule can be approximately equated to a unit per year. Generally the NDE was completed in the summer and the ECA in the winter. The results to date have shown, that apart from a few defects which have been rectified, the high pressure components are acceptable for another 25 years service.

➤ Continued safe and reliable operation of Ffestiniog for a further 25 years is dependant on the integrity of the major high pressure components especially embedded parts.
➤ Scope: Turbine spiral casings, pump spiral casings, main inlet and Pump discharge valves (See figure 3).

COMPONENT REVIEW OBJECTIVES

➤ The primary objective of the review of the high pressure components is to demonstrate, that the plant can be safely operated for the next 25 years
➤ 25 years future operation equates to approximately 75,000 turbine pressurisations
➤ At the end of the 25 years the components should still be able to withstand the worst case over pressurisation ~ equivalent to a 90 MW trip

PHILOSOPHY AND STAGES FOR REVIEW OF HIGH PRESSURE COMPONENTS

Accurately quantifying the remaining life and condition of a component or plant can only be achieved by the completion of an Engineering Critical Assessment. The ECA analyses the components / plant, to evaluate its future fitness for purpose for its given predicted operation and environment. This evaluation is carried out by examining the structure with respect to; plastic collapse, fracture mechanics, fatigue, corrosion, erosion and wear.

To facilitate a ECA the review was broken down into three key stages;

Stage 1   Collection of known data ~ original drawings, historical loading, future loading, modifications etc. The research and collection of the above information can prove challenging particularly historical data, which may not be available or archived.

Stage 2   Non-destructive examination (NDE) of the components by Ultrasonic (UT) and Magnetic Particle (MPI) techniques. Thickness and geometric checks were also carried out.

Stage 3   Examination of the components via an Engineering Critical Assessment (ECA). The integration of data from the first two stages along with detailed finite element analyses results allows assessments to be made of the components. The assessment is split into two distinct areas; Firstly the analysis of known defects and secondly the analysis of the remaining structure. Once this has been completed accurate and safe conclusions on components remaining life can be reached. If the component fails the assessment acceptance criteria then remedial action would be required.

REVIEW RESULTS TO DATE

As Ffestiniog is a working station the NDE has been carried out when dewatered outages were available. The corresponding schedule can be approximately equated to a unit per year. Generally the NDE was completed in the summer and the ECA in the winter. The results to date have shown, that apart from a few defects which have been rectified, the high pressure components are acceptable for another 25 years service.

UNITS 3 AND 4 MANDOOR DEFECTS

During the ECA of units 3 and 4 turbine defects propagating from the mandoor area were found to be unacceptable in their present condition. PD 6493 uses Failure Assessment Diagrams (FAD) to assess the acceptability of defects. The following slides shows the defects position on the FAD.

The two defects were found propagating from the manhole hinge cut-out. The manhole is one of the most highly stressed areas of the spiral structure. The defects are located in the corners of the manhole. The corner surfaces appeared to be a flame cut finish as specified in the original drawings. The defects were corner flaws having the dimensions, Manhole defect north 50 mm x 20 mm, Manhole defect south 20 mm x 18 mm. The following slides show defect location and sizes (See figure 4).

Due to the amount of unknowns in the assessment a high level of conservatism was present in the assessment. To refine the assessment further and remove some of the unknowns, the following activities were carried out:

1   The removal of turbine spiral material, for testing of physical and mechanical properties including fracture toughness. It was found that the fracture toughness property was double that predicted from the failed unit material testing.
2   Strain gauge measurement on a non-cracked unit around the mandoor area to validate the finite element analysis in this area. The strain gauge measurements showed that the finite element analysis was accurate with yield stresses in the corners.
3   A more detailed finite element analysis of the mandoor area was carried out.
4   1 Re-examination of the defects using MPI, UT and Alternating Current Potential Drop (ACPD) techniques. Using ACPD the manhole defects were found to be larger than predicted by UT. This moved the assessment point towards the boundary between the acceptable and unacceptable zones.

A detailed Risk Assessment was carried out to evaluate the consequences of one of the mandoor defects causing catastrophic failure of the turbine spiral and possibly flooding the station. It concluded that the concrete local to the spiral would play a considerable part in minimising the flooding potential that existed. It was also decided that a programme of monitoring for any change in the defect geometry be adopted. The following techniques were adopted for monitoring any change in defect geometry; magnetic particle, ultrasonic, alternating current potential drop and acoustic emission monitoring.

The re-assessment of the defects concluded that the mandoor defects were safe but any growth would result in them being unacceptable. See figure 5 for re-assessed FAD. The ECA predicted that the defects would grow to unacceptable levels within a very short period.

THE REPAIR

With the defects being close to unacceptable levels, there was a need for some form of rectification work to be carried out. Many different options were evaluated, ranging from crack stitching, to removal and replacement of the mandoor area. The final preferred option was, removal of the defect followed by welding, as this was the most effective solution from a strength and unit outage period perspective. However, this was not without its complications; distortions, inducement of defects and material weldability factors were carefully examined. A temper bead welding technique was proposed for the repair. This technique has been established for many years as a mechanism for HAZ microstructure modifications and avoidance of post weld heat treatment. The repair work, was also carried out on the large surface breaking defect, this repair was successfully carried out on site with the minimal outage time. The units were brought back on-line successfully and the structural integrity of the welds were shown to be sound by subsequent NDE. The re-assessed mandoor area gave a predicted life of over 75,000 turbine pressurisations equivalent to 25 years life.

CONCLUSION

The study and remedial work carried out for the high pressure components has proven that the components life at Ffestiniog Power Station will be greater than 75,000 pressurisations (25 years). To achieve this the components will require to be continuously inspected and maintained as they have been in the past. The inspection requirements having been defined by the ECA.

The review of high pressure components utilising a ECA is a logical and proven method of assessing future life of a component. However such a task can only be accomplished by an established method of working involving equipment suppliers and the plant operators. The work carried out in Ffestiniog demonstrates that teamwork and co-ordination between turbine designers and the plant operator can provide a substantial contribution towards asset management and provides continued confidence in the operation of ageing plant.

Future operating decisions, which effect the high pressure components, can be made with confidence now the condition of the major components is known.

- Situated in North Wales

- First major pump storage scheme in Great Britain

- Designed in mid to late 50's

- 4 Vertical 90 MW Units

- Francis turbines, Main Inlet Valves and Pump Discharge valves supplied by English Electric

- Vertical two-stage pumps supplied by Sulzer Brothers

- Started operating in 1963

- Over 90,000 turbine pressurisation's per unit to date

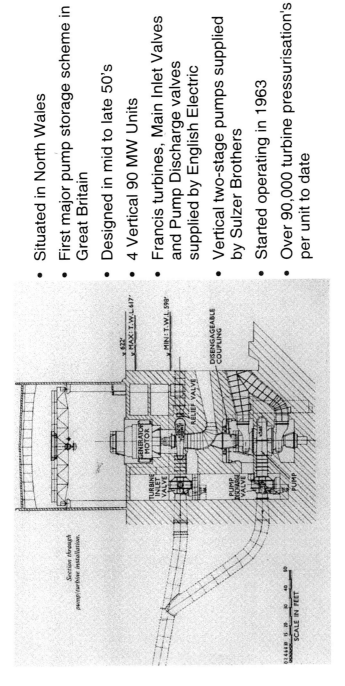

Figure 1 section though a unit at Ffestiniog

Turbine spiral unit 3 pressure test failure (1960) at English Electric workshops

- Pre-existing crack of 203 mm from a weld repair of a 1371 mm crack in the parent plate material BS968: 1941.

- Coarse grain size & laminations / inclusions found in parent plate

- Low impact strength of parent plate

- Weldability of plate poor over 24.5 mm thickness (plate thickness ranges from 60 - 24.5 mm)

- Problems with this unit during manufacture

- Last unit produced other units embedded at site

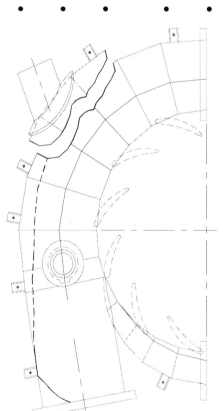

Figure 2 brittle fracture of unit 3 turbine casing after hydraulic pressure test in 1960

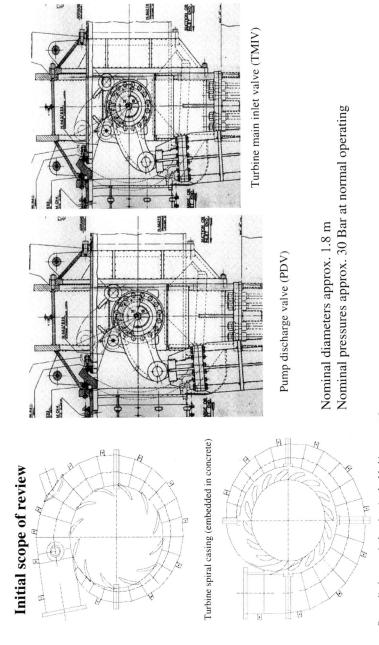

Initial scope of review

Turbine spiral casing (embedded in concrete)

Pump discharge spiral casing (embedded in concrete)

Pump discharge valve (PDV)

Turbine main inlet valve (TMIV)

Nominal diameters approx. 1.8 m
Nominal pressures approx. 30 Bar at normal operating

Figure 3 scope of high pressure component review

Turbine unit 4 critical defects

Manhole - Defects in corner cut-outs

Corners flame cut finish

Outside

Inside

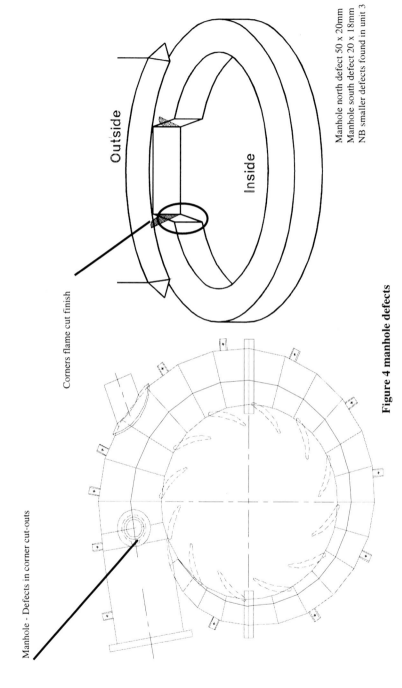

Manhole north defect 50 x 20mm
Manhole south defect 20 x 18mm
NB smaller defects found in unit 3

Figure 4 manhole defects

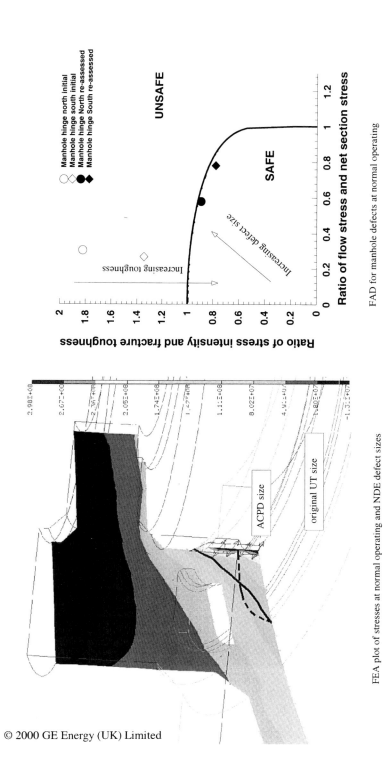

FEA plot of stresses at normal operating and NDE defect sizes

FAD for manhole defects at normal operating

Figure 5 FEA of manhole and FAD of manhole defects

© 2000 GE Energy (UK) Limited

S724/005/2000

Generator motor air cooler improvements within hydroelectric power generation

D J CARSON-MEE and **W O MOSS**
Edison Mission Energy, First Hydro Company, Llanberis, UK

ABSTRACT

Edison Mission Energy own and operate Ffestiniog and Dinorwig pumped storage power stations with a combined output of 2200MW. After 15 years of operation at Dinorwig power station, the existing spiral wound generator motor heat exchangers had degenerated due to a mix of stress corrosion and fatigue at the tube plates. Renewal of the heat exchangers has not only improved the design but has increased the thermal transfer rates and consequently reduced the generator temperatures.

Heat transfer within the power industry is paramount to the longevity of its equipment. It has been evaluated that a ten degree centigrade temperature increase on the generator motor windings will present a significant reduction in generator life[1]. Within Dinorwig power station, the generator motor core winding temperatures have been conservatively reduced by ten degrees on load by introducing new plate fin heat exchangers. The plate fin heat exchangers have also demonstrated better resistance to fin damage, improved vibration characteristics, quick assembly times and have reduced fouling rates. Edison Mission Energy, First Hydro have now extended the use of such plate fin heat exchangers in Ffestiniog power station. Innovative fin designs are also being trialled, that have been demonstrated to improve heat transfer rates by a further 8-10% from the plate fin designs.

This paper looks at a hydro electric industrial application of such technology and the benefits that have been made because of the modifications and also examines the future of improved heat exchanger designs within the power industry.

1. INTRODUCTION

The generator motor at Dinorwig Power Station is air cooled by a closed loop ventilation system. The heat is transferred from the chamber by 8 water cooled heat exchangers. The heat exchangers are fed with water from the station cooling water system. Recent development work on the generator motor has necessitated the installation of new heat exchangers. The existing designs were inherent to premature failure and were demonstrated to be of poor heat exchange capacity. Tests have demonstrated that the plate fin design were superior from both of these aspects and have therefore been specified for installation into all six units. Temperature drops of between 8-10°C have been witnessed on the stator core temperatures when at full load, representing a 6.1% improvement in thermal transfer rates. A new cooler design with improved fin designs is currently under trial, and preliminary indications would show a further improvement of about 8.7% on thermal transfer rates, which could have a further reduction of up to 8°C on the rotor

2. OBJECTIVES

The principle aim of changing the generator motor coolers was to increase the thermal performance of the coolers and their reliability. The installation of the new coolers would also reduce machine downtime through cooler failure and increase the thermal cooling content of the generator motor. Failure of the coolers can result in water ingression to other components within the generator motor, which could lead to catastrophic consequences. A general arrangement of the generator motor is shown in *Figure 1*.

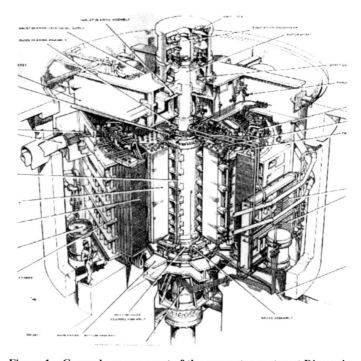

Figure 1 – General arrangement of the generator motor at Dinorwig

3. SYSTEM CONFIGURATION

The cooling water system for Dinorwig Power Station originates from the lower lake llyn Peris. This is transported 80m underground and from here branches to appropriate areas within the station. Each generator motor has a separate feed that runs in a ring main manner within each generator motor enclosure, circulating through and out of the 8 heat exchangers. Air is forced past the stator windings and through the air coolers using 6 axial and 12 vertically mounted fans as shown in *Figures 2 & 3*. There are 8 coolers per unit, each carrying approximately 16.3litres per second of cooling water in a cross/partial counter flow configuration. Approximately $18.4m^3$ per second of air enters each cooler.

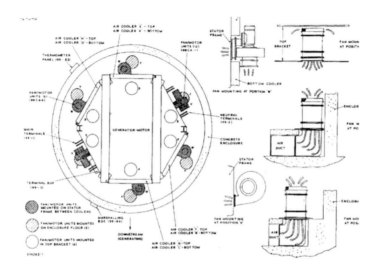

Figure 2 – Air circulation inside the Generator Motor

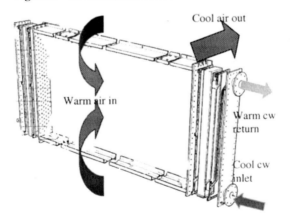

Figure 3 – Typical generator motor heat exchanger

4. COOLER TRIALS

4.1 Original Design

The original system configuration was the same as discussed in *section 3*, with a designed cooling air capacity from 61.4°C to 40°C at 18.4m^3 per second, representing a heat load of approximately 475kW. The original design consisted of spiral wound, tin coated, copper fin coolers. The premature failure of this type of cooler was associated with tube fretting around the brazed areas at the tube sheet. This was associated to the coolers resonance and inadequate support of the tubes. A typical tube failure can be seen in *figure 4*. The design also presented poor flow paths, enabling dirt and ingress to build up on the fins quickly. There were also areas where the fins did not cover near to the tube sheets, consequently reducing the available heat transfer area.

Figure 4 – Typical tube fracture, shown plugged

4.2 Trials

Because the life expectancy of the existing units was becoming critical to the operation of the station, trials were conducted with two different designs of heat exchanger. The designs employed were an improved spiral wound unit and a plate fin unit, sourced from three suppliers. The coolers were also specified to increase the thermal efficiency, improve vibrations and fin fouling. The trials lasted 6 months, from which water and air temperatures, pressures, vibrations, fouling and general reliability checks were monitored. The tests confirmed that the plate fin block variety showed the best performance and were therefore chosen.

5. SERVICE RECORD

5.1 Generator Motor Temperatures

The plate fin block coolers chosen utilised the same system configuration as discussed in *section 3*. The heat exchangers incorporated 162 × 3.2m, 5/8inch admiralty brass tubes, fitted with 29mm wide aluminium fins spaced at 11 fins per inch, see *Figure 5*.

S724/005/2000 © IMechE 2000

Figure 5 - Plate Fin Coolers

The cooling capacity was rated from 61.4°C to 38.9°C, representing a heat load of approximately 504kW[2]. The findings showed that the fin block coolers were able to carry a larger heat transfer, offering air temperature reductions of up to 45°C at maximum loading conditions on the machines, see *figure 6*.

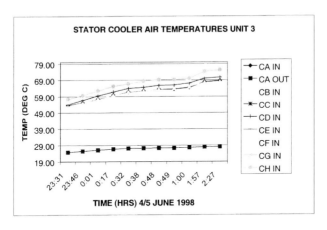

Figure 6 – Air temperatures across the cooler.

The thermal increase represented a 6.1% improvement from the original spiral designs, which translated into a 10°C temperature reduction of the generator motor at full load, see *Figure 7*.

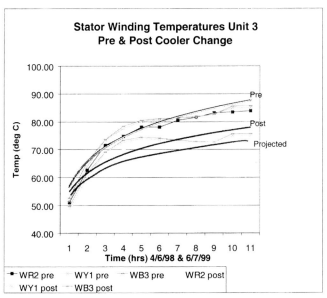

Figure 7 – Stator temperatures pre, post & projected.

5.2 Future Development

A new fin design is currently under trial for a single unit at Dinorwig. Preliminary indications from the tests indicate that the cooling capacity would be from 61.4°C to 36.9°C representing a heat load of 548kW[3]. This represents a further improvement of 8.7%, which could lead to a conservative generator motor temperature reduction of up to 8°C as shown projected in *Figure 7*.

6. CONCLUSIONS

The plate fin block coolers have shown their advantages for generator motor applications in arduous running environments such as those presented by Dinorwig Power Station. The improvements made by the first phase of plate fin design have shown temperature drops in excess 12% on the generator temperatures. With the advances of the improved plate fin design, the stator core temperatures have been projected to fall by a further 10%. This would represent an increased life expectancy of the generator motor, though exactly by how much is a function of the running hours on condition. The benefits gained in increasing the life expectancy of the generator motor and reducing unit downtime far exceed the outlay of the coolers.

It is anticipated that the first actual cooler unit with the new fin design will be installed and monitored at Dinorwig sometime during December 2000. Results that verify the tests shall be available shortly following load conditions for that unit.

REFERENCES

[1] BS2757(1986), IEC85(1984) – Determining the thermal classification of electrical insulation – British Standards Institution.

[2] Britannia Heat Transfer (1997) – Elfin plate fin design calculations – *Appendix D to design specification for FHC.*

[3] Jones, H (2000) – Generator Motor Stator Core Coolers Performance – *Internal communication, BHT/FHC.*

[4] Fenwick, GT (1996) – Generator air cooler design for a powerplant refurbishment – *Hydropower & Dams, Issue 5.*

[5] Rogers & Mayhew (1983) – Engineering thermodynamics work and heat transfer – *Longman Group Ltd, 3rd edition*

New Projects

S724/006/2000

Design and supply of mechanical equipment for the Muela Power Station, Lesotho

G C JONES and **R G TAYLOR**
GE Energy (UK) Limited, Doncaster, UK

1. INTRODUCTION

The Lesotho Highlands Water Project is a multi purpose project aimed at developing the water resources of the mountainous region of Lesotho through a series of dams, tunnels, pumping stations and a hydro electric facility. The primary objectives of the project are to:

- Redirect some of the water flowing out of Lesotho towards population centres in South Africa
- Provide water supply, irrigation and regional development in Lesotho
- Generate hydro electric power utilising the redirected water.

This paper is focused on the equipment supplied for the power generation which is housed in an underground power station located at the end of a 45km water transfer tunnel extending from the Katse reservoir. The 72MW power plant discharges into a regulating pond which in turn leads to a further 37km discharge tunnel with an outfall in South Africa.

Attachment 1 provides a geographical overview of the project location.

GE Energy UK have had extensive involvement in the Project, supplying and delivering a variety of equipment for the project in two roles:

1) In consortium with ABB of Sweden, where GE was the main contractor for the supply of the Powerhouse mechanical equipment.

2) As a Sub -Contractor to the Civil Contractor on the Station By pass system where GE supplied the protection and dissipating valve.

The paper being presented today looks at the first of these roles. The contract was awarded to the E&M consortium in September 1993, with the contract becoming effective on the 1st April 1994. In general ABB supplied all the electrical equipment, with GE Energy (UK)

Limited supplying the mechanical equipment, as well as the heating and ventilation system. The balance of building services and the station cranes were supplied as part of the civil contract.

The project structure for the Powerhouse, comprised of Lesotho Highlands Development Authority (LHDA) as the client, supported by their consultant the Lahmeyer Macdonald Consortium (LMC). Muela Hydro Power Contractors (MHPC) undertook the civil works.

This paper will briefly discuss the GE Energy equipment supplied for the Power plant noting the more significant aspects of the supply.

2.   FRANCIS TURBINES

2.1   General

The three hydraulic turbines were designed and supplied by GE Energy from the United Kingdom. They are of a conventional vertical Francis reaction type with the turbine shaft coupled through an intermediate shaft to the generator rotor. Table 1 provides some of the turbine technical particulars.

Table 1 Technical Particulars	
Maximum gross head	293m
Minimum gross head	229m
Rated output at 236m head	25.2 MW
Runner throat diameter	1.085m
Number of runner blades	17
Turbine speed	750 rpm
Turbine design pressure	36 Bar
Number of guide vanes	20

The water on the project exhibits low scale forming tendency resulting in a relatively high corrosivity which presented some challenges in the selection of coatings and materials.

2.2   Hydraulic Design and Model Testing

The runner for Muela was selected from the library of turbines available within GE Hydro. The design was typical for a medium to high head Francis machine. The operating range of the upstream reservoir is in excess of 60m, which together with significant friction losses in the 45km long headrace tunnel, resulted in a relatively large net head range for turbine operation. This demanded a runner which was capable of smooth operation at unit speeds (n11) well below the unit speed for best efficiency.

A witnessed homologous model test of the turbine was carried out in December 1995 at the GE Energy Laboratory in Sweden. Its purpose was to demonstrate the performance of the model turbine and to ascertain that the guaranteed weighted efficiency and other performance characteristics could be met.

2.3 Runner and Shaft

The runner for Muela was machined from a one-piece 17/4 chrome nickel alloy casting. The material was specified to address the water corrosivity. A fabricated stainless steel nose cone was welded to the runner before final balancing. The flange connection between the runner and the forged turbine shaft consists of shoulder bolts to transmit axial thrust and five shear bushes for the torsional load. The intermediate shaft connection to the turbine shaft has fitted bolts.

2.4 Spiral Casing and Guide Vane operating gear

The fabricated carbon steel spiral casing and stayring assembly was completely welded and pressure tested to 1½ times working pressure at works.

For site embedment, a partially pressurised technique was adopted with the spiral casing maintained at 2.4 MPa during the second stage concreting.

The top and bottom covers are fabricated from carbon steel plate. The 20 guide vane bearing housings in the top cover are fitted with self-lubricating bearing sleeves. For the lower bearings, the sleeves are inserted directly into the bottom cover. Guide vane height is set using an adjustment screw from below, which also serves to support the guide vane axial thrust. The guide vanes are machined from cast 13-1 Cr Ni stainless steel. The externally mounted regulating ring operates the guide vanes through the links and levers, which are connected to the vanes by a friction device. This protects the turbine in the event of obstruction to closure of guide vane. The two servomotors for the regulating ring are mounted in the turbine pit.

A shop assembly of the main turbine stationary parts was carried out to check the operation of the guide vane operating gear as well as verification of dimensions and concentricity.

The enclosed turbine pit has access from two sides, which allows for upwards and sideways dismantling of the top cover, rotating parts and bottom cover without disturbing the generator.

Attachment 2 shows a cross section through the turbine.

2.5 Shaft Seal and Turbine Guide Bearing

This is a mechanical seal of the axial type, comprising a composite seal face acting on a corrosion resistant cone mounted on the turbine shaft. Seal contact pressure is hydraulically balanced. An inflatable static seal is also provided.

The guide bearing is a well-proven tilting pad design. The total oil volume is contained within the sump and is self-circulating. Water/oil coolers are mounted within the sump. Connections are provided for oil purification as well as all the necessary instrumentation to monitor oil levels and temperature.

3. GOVERNING SYSTEM

The electronic governor supplied by GE Energy is an ABB type HPC 640, incorporating digital technology. The basic features of the governor are power and guide vane control, frequency control, start/stop logic, limiter functions, automatic synchronising, operational mode detection and automatic changeover of governor parameters, monitoring functions and man/machine communication. The governor supplied also includes software for surge tank compensation, to reduce surge tank oscillations.

The Governor Electro Hydraulic Control Unit, of GE Energy design, is based on modern fluid power technology, with a proportional valve as the interface between the electronic governor and the Hydraulics. The hydraulic pumping unit and air/oil receiver provides the governor oil system pressure at 42 bar. The unique feature is the manual control, independent of the electronics and the proportional valve. A turbine overspeed device is also provided which operates a plunger operated hydraulic control valve, which in turn initiates a closure of the guide vane servomotors. A dual closing feature of the guide vane servomotor ensures that the speed rise and pressure rise, due to load rejection, is within the guaranteed values.

4. TURBINE INLET VALVES

The three 1.15m turbine main inlet valves are of a rotary spherical type of cast steel construction. Although normally operating under balanced pressure conditions, the valves can in an emergency close against a burst flow of 45m³/s. This GE design of valve provides high reliability as well as rugged construction to meet demanding operating conditions.

Valve sealing is achieved by means of a sliding service seal ring, which is operated by penstock pressure. Operation of the service seal is automatic and part of the normal valve operation sequence. At the upstream end of the valve a maintenance seal is provided, which is applied, using a hand pump, for high integrity isolation.

The movement of the inlet valve rotor is by the action of a single double acting servomotor operating under hydraulic oil pressure. The necessary interlocks between the rotor movement and service seal operation is achieved by mechanically actuated control valves operated by the rotor lever arm. The operation of the inlet valve is also hydraulically interlocked with the draft tube flap gate.

The hydraulic oil system for the inlet valve has an operating pressure of 120 bar supplied from a pumping unit combined with a nitrogen piston accumulator system. An offline filter system is provided to maintain the oil at a low contamination level.

Attachment 3 provides a section through the Inlet Valve.

S724/006/2000

5. DRAFT TUBE FLAP GATE

Isolation of each turbine from the tailrace for maintenance is achieved using a flap gate mounted in each of the three draft tubes. The flap gates are a fabricated construction 2.65m span x 1.42m deep to GE Energy design. These gates are normally open and only closed when it is necessary to dewater the respective turbine. When fully open the gate is in-line with the roof of the draft tube providing no obstruction to flow. The gate and its bonnet assembly are designed for complete removal if required for repair and maintenance.

Operation of each gate is by two servomotors complete with a dedicated power unit and nitrogen accumulator system. The hydraulic system pressure is 120 bar. The operation of the flap gate is hydraulically interlocked with the turbine inlet valve. Gate operating torque is transmitted by way of torque arms and shafts passing through the bonnet assembly with self-lubricating bearings and seals. A positive drive connection between the arm and the shaft is achieved using taper drive keys. A particular feature of the gate design is that the torque shafts allow the servomotors to be mounted within the power plant together with the associated control system.

Attachment 4 provides a section through the Turbine, Inlet valve and Flap gate.

6. PENSTOCK GUARD VALVE

The 2.5m penstock protection valve is located in the guard valve chamber at the top of the vertical steel lined penstock, downstream of the surge chamber. This valve is designed to isolate the lower penstock and to close under emergency (burst) conditions.

The valve construction is based on a well proven GE Energy design comprising of a double seal butterfly type lattice disc within a one piece cast steel body. The double seal lattice allows a maintenance seal to be applied to enable inspection or replacement of the downstream seal. Opening of the valve occurs under balanced pressure conditions under the action of two single acting hydraulic cylinders supplied with pressure oil from a dedicated pumping unit and accumulator. The valve closes under the influence of weighted levers and utilising the cylinders as dashpots. Closing tendency is also assisted by eccentricity in the disc. Pipework and needle valves are provided to allow balanced conditions to be achieved before the valve can be opened. Anti-vacuum valves are mounted downstream of the valve for air admission during closure. Over-velocity in the pipeline is detected by a trip switch, which initiates the closure of the protection valve.

7. COMPRESSED AIR SYSTEMS

GE Energy supplied two compressed air systems for the power plant. A high-pressure system is required to charge and automatically top up the air/oil receivers for the governor system. For the generator brakes and the station air supply a low-pressure system is required. Both systems are separate and completely independent.

8. WATER DISTRIBUTION SYSTEM

The water distribution system at Muela is a complicated system which is used to provide potable water, as well as raw water for fire fighting purposes and an emergency supply for the unit cooling water. Water at penstock pressure is supplied from a tapping upstream of the penstock guard valve to replenish the surface raw water reservoir situated on a hill above the power plant. This reservoir supplies the emergency supplies as well as a water treatment unit situated in the workshop near the operations building. Filtered and treated water is then distributed to the underground powerhouse, operations building and to a local village supply tank.

9. COOLING WATER SYSTEM

Cooling water for the turbine, generator, transformers and HVAC system within the powerhouse is provided as a separate system. The system is pumped from Unit 3 draft tube and comprises duty pumps, auxiliary pumps and main strainers. Two additional pumps boost the cooling water pressure to the shaft seal. A cyclone separator and a fine strainer ensure contamination level at the shaft seal face is limited. The discharge from the cooling water consumers is to the tailrace tunnel.

Emergency cooling water to the shaft seals in case of pump failure can be provided from either the pressure tank or operated manually from the water distribution system.

10. DRAINAGE & DEWATERING SYSTEM

As with all underground power stations, the integrity of the plant is dependant on a reliable drainage and dewatering system.
The main drainage system at Muela comprises a main station sump with three submersible pumps of nominal capacity 245 l/s discharging to the tailrace. A smaller auxiliary sump and associated pumps drain leakage water from the dewatering gallery into the main sump. The main sump also includes an eductor motivated by penstock pressure for service in the event of pump power failure. All the pumps and the eductor are operated automatically under level switch control.

Dewatering of the penstock and turbine is achieved through a separate dewatering sump. Two vertical turbine pumps operate under level switch control and discharge outside the station. The dewatering sump can also be utilised for emergency station drainage.

11. OIL HANDLING SYSTEM

The bulk oil handling for the turbine and generator bearings is provided. The system comprising storage tanks purifier, transfer pumps and associated pipework linking the units and is situated in a separate room at the turbine floor of the powerhouse.

12. HVAC EQUIPMENT

The HVAC system provides heating and ventilation to the underground cavern. The scope of works included the supply of the air handling unit, pressurisation fans, ductwork, grills, fire dampers, smoke dampers and controls.

The air-handling unit located in the crown of the cavern draws fresh air from the surface and then distributes the air, via a system of ducts, to the various floors within the powerhouse. The distributed air returns to the surface via either the main access tunnel or cable shaft. The system can operate in winter or summer modes. The system is also designed to respond to a fire in the cavern by shutting down the air handling unit and opening smoke dampers in the roof to exhaust smoke through the cable riser.

13. SITE INSTALLATION AND COMMISSIONING

Site mobilisation for the E&M works at Muela started in August 95, however due to a strike by the civil work force the first draft tube was not installed until January 1997, with unit 1 spiral casing installation and pressure test following five months later. The erection work was carried out by predominantly local and South African labour under GE Energy supervision. Unit 1 was successfully commissioned and taken over in September 98, with the other two units following at approximately two month intervals. The completion of the work was in accordance with the client's programme, despite the early delays and civil unrest within Lesotho.

Attachment 5 demonstrates the overall Muela site programme

14. PERFORMANCE TESTING

All three turbines were index tested by GE Energy during January 1999. Due to the good co-relation across the results from all three units, LHDA decided to cancel the requirement to carry out a full site efficiency test.

15. TRAINING

Under the contract with LHDA, GE Energy was required to provide a structured training programme for the station operations and maintenance staff. This was provided for three groups of ten persons. Training period for each group was four weeks. Classroom lectures, works and site visits were arranged as well as a visit to a hydraulic laboratory. The courses were concluded with a short examination and assessment.

Attachment 6 provides a typical training programme.

16.   CAVITATION INSPECTION

A contractual cavitation inspection of the Muela turbines was carried out in May 2000. Operating period for the three units at that time varied between 8500 and 10000 hours. No serious pitting was evident on any of the units with the only visible trace of cavitation being limited to a slight greying on the suction side of some of the blades and in the vicinity of the unloading holes. The damage was well within the guaranteed values to IEC 609 and accepted by the client.

Muela Project Location – Attachment 1

GE Energy (UK) Limited

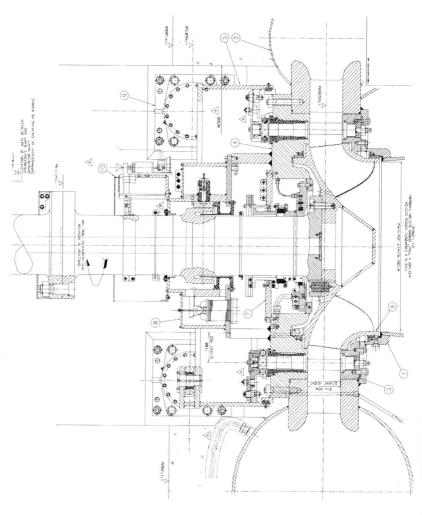

Turbine Cross Section – Attachment 2

GE Energy (UK) Limited

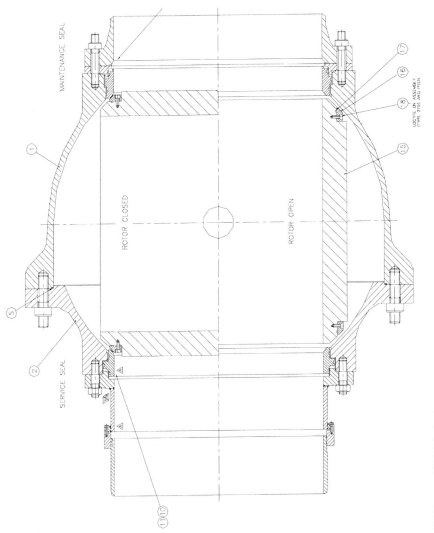

MAINTENANCE SEAL

ROTOR CLOSED

ROTOR OPEN

SERVICE SEAL

① ⑤ ② ⑪⑩ ⑰ ⑯ ⑧ ⑮

LOCTITE ON ASSEMBLY
(TYPE 270) AND PEEN

Section through Main Inlet Valve – Attachment 3

GE Energy (UK) Limited

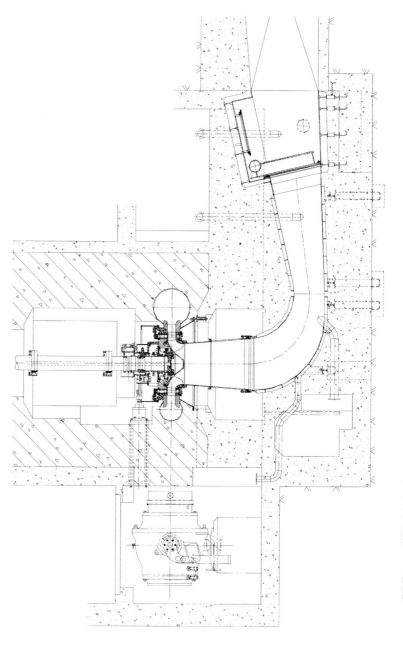

Cross Section of Turbine, Inlet valve and Flap Gate - Attachment 4

GE Energy (UK) Limited

S724/006/2000

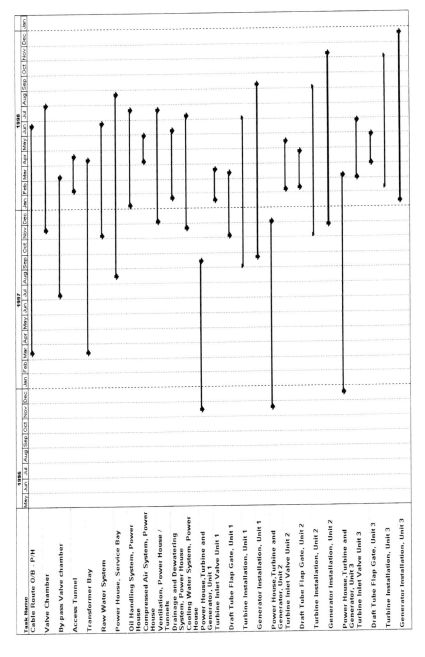

Muela Site Programme – Attachment 5

GE Energy (UK) Limited

Task Name
Cable Route O/B - P/H
Valve Chamber
By-pass Valve chamber
Access Tunnel
Transformer Bay
Raw Water System
Power House, Service Bay
Oil Handling System, Power House
Compressed Air System, Power House
Ventilation, Power House / Tunnels
Drainage and Dewatering System, Power House
Cooling Water System, Power House
Power House,Turbine and Generator, Unit 1
Turbine Inlet Valve Unit 1
Draft Tube Flap Gate, Unit 1
Turbine Installation, Unit 1
Generator Installation, Unit 1
Power House,Turbine and Generator, Unit 2
Turbine Inlet Valve Unit 2
Draft Tube Flap Gate, Unit 2
Turbine Installation, Unit 2
Generator Installation, Unit 2
Power House,Turbine and Generator, Unit 3
Turbine Inlet Valve Unit 3
Draft Tube Flap Gate, Unit 3
Turbine Installation, Unit 3
Generator Installation, Unit 3

Typical Training Programme – Attachment 6

Topic	Day	9.00-10.30	Break	10.30-11.45	11.45-12.30	Lunch	1.30-2.30	Break	2.45-3.30	3.30-4.00
Welcome To GEUK & Introduction to Muela	1	Domestic Arrangements RGT		Introduction to GEUK DES	Introduction to Muela RGT		Introduction to Muela RGT		Routine Maintenance RGT	Meet the Lecturers of GEUK
Francis Turbine	2	Hydraulic Design LIM		Introduction to Mech Design N/W	Turbine G.A MAM		Embedded Parts MAM		Routine Maintenance MAM	Questions & Answers
Governing System	3	Introduction to Muela Governor JLK		Overview of Governor System JLK	Components of the Governor System JLK		Components of the Governor System JLK		Components of the Governor System JLK	Questions & Answers
Turbine Inlet Valve / Penstock Guard Valve	4	Valve Principles HTB		Turbine Valve	Penstock Protection Valve		Operation of Valves of JLK		Operation & Maintenance of Valves JLK	Questions & Answers
Draft Tube Flap Gate	5	Gate Principles M.H		Features of the the Muela Draft Tube Flap Gates M.H	Features of the the Muela Draft Tube Flap Gates M.H		Control of Gate JLK		Maintenance of Gate M.H	Questions & Answers
Compressed Air System	6	Introduction to to H P & L P Air System JLK		Features of & layout of the equipment for the L P air System JLK	Features of & layout of the equipment for the H P air System JLK		Operation of L P & H P Air System JLK		Maintenance of L P & H P Air System JLK	Questions & Answers
Water Distribution System	7	Introduction to Water Distribution System JLK		Features of & layout of the water dist sys JLK	Features of & layout of the water dist sys JLK		Operation of the water Dist System JLK		Maintenance of water Dist System JLK	Questions & Answers
Drainage & Dewatering System	8	Introduction to D/DW System JLK		Features of D/DW sys JLK	Features of D/DW sys JLK		Operation of D/DW System JLK		Maintenance of D/DW System JLK	Questions & Answers
Water Measuring System	9	Introduction to D/DW System JLK		Features D/DW sys JLK	Features D/DW sys JLK		Operation of D/DW System JLK		Maintenance of D/DW System JLK	Questions & Answers
Oil Purification System	10	Introduction to Oil Purification System JLK		Features of Oil Purification System JLK	Features of Oil Purification System JLK		Operation Oil Purification System JLK		Maintenance Oil Purification System JLK	Questions & Answers
Mechanical Workshop	11	Introduction to Mech Workshop JLK		Visit Dunsley	Visit Dunsley		Visit Dunsley		Visit Dunsley	Questions & Answers
Heating and Ventilating	12	Introduction to H V A C N K		Features of HVAC JLK	Features of HVAC JLK		Operation HVAC JLK		Operation HVAC JLK	Questions & Answers
Operation and Maintenance	13	Introduce O & M Manuals SMS		The use of O&M Manuals SMS	The use of O&M Manuals SMS		The use of O&M Manuals SMS		The importance of reg maintenance SMS	Questions & Answers
Sheffield University	14									Questions & Answers
Visit to U.K power Station	15									Questions & Answers

GE Energy (UK) Limited

© 2000 GE Energy (UK) Limited

Gitaru Unit No. 1 – 80 MW in three years

F GRIFFIN and **A YOUNG**
Mott MacDonald, UK
A WEBB
Formerly with Mott MacDonald, UK

BACKGROUND

In October 1996 Mott MacDonald with Knight Piésold were appointed by the Kenya Electricity Generating Company Ltd (KenGen) to engineer the installation of the third and final unit (Unit 1) within the Gitaru power station situated on the Tana River approximately 160 km north east of Nairobi. The first two units at Gitaru were commissioned in 1978 and at that time it was expected that the third machine would follow after two or three years. However, it was not until 1996 that KenGen decided to proceed with implementation as it was decided that this machine was the most cost-effective means of quickly overcoming a lack of generating capacity.

Our work was to be carried out in two stages. Stage I would review the technical design of the scheme to confirm or otherwise the original design parameters and additionally determine the best contractual arrangements for implementation. Stage II would include the preparation of technical specifications, tendering, contract award, design review and construction supervision.

STAGE I

Design Review
The original proposals anticipated that the third machine would be similar if not identical to the first two. However, due to the long delay it was considered to be appropriate to review the overall specification for the new machine. Two of the more important changes are discussed below.

Turbine-Generator Rating
The original machines were each rated at 72 MW. Studies showed that the marginal costs of increasing capacity were relatively low and additional capacity would allow more energy to be generated when river flows were high. However, as all three machines share a common tailrace tunnel, larger turbines would produce greater flows with subsequently higher friction

losses. Transient surge and water hammer effects would also be more significant. It was decided to increase the unit rating such that with all three machines on full load the maximum flow capacity of the tailrace tunnel was achieved. This fixed the rating of the new machine at 80 MW.

Connection to HV Transmission System

The original design expected the third machine to connect into the 132 kV switchyard at Kamburu, 9 km from Gitaru. Subsequently, the transmission system has been expanded and a 220 kV network now forms the backbone of the Kenyan electrical grid. System studies demonstrated that under certain circumstances power transfer restrictions between the 132 kV and 220 kV switchyards at Kamburu would restrict generating capacity. This limitation as well as the lower in-service losses of 220 kV equipment offset the additional capital costs of higher voltage equipment and it was decided that the new machine should be directly connected to the 220 kV switchyard.

The 220 kV switchyard is of the "mesh" type which is a cost-effective arrangement that is well suited to transmission substations where switching usually only takes place following faults or during maintenance. However, when feeders are connected the mesh has to be opened and a single fault would lead to load being disconnected. If Gitaru Unit 1 was synchronised within the Kamburu 220 kV switchyard the mesh would need to be opened every time a machine was started or shutdown and this would significantly reduce the reliability of this important grid switching station. It was therefore decided that the new transmission line and generator transformer should normally remain energised and Unit 1 should be synchronised at a new circuit breaker at the generator 15 kV terminals. This arrangement has the secondary advantage of allowing station auxiliary power supplies to be supplied directly from the 220 kV system even when all three machines are shutdown.

Project Scope

The intake gate, inlet penstock and draft tube and draft tube gate were already installed and many provisions had been made in the design of the power station electrical and mechanical services for the addition of the third machine. It was therefore a relatively straightforward job to install the new machine. The final scope of the project was as follows:

- 81 MW Francis turbine, inlet valve and associated control systems.
- 95 MVA generator and auxiliaries.
- Additional set (standby) of 110 V and 48 V d.c. battery equipment.
- 95 MVA 15/220 kV generator transformer with on-load tap changer.
- 15 kV circuit breaker and outgoing cable connections to transformer.
- Fully automatic control system with provision for full remote control and incorporation into national SCADA system.
- Switchyard equipment at Gitaru.
- 9 km single circuit 220 kV transmission line.
- Additional bay of 220 kV switchgear at Kamburu power station.
- All civil works associated with the above.

Contract Arrangements

KenGen had two main requirements; firstly the project should be implemented in the shortest possible time and secondly that costs should be minimised. We examined how the contracts should be structured to best achieve these somewhat conflicting requirements. In order to minimise the time preparing specifications and tendering it was agreed at an early stage that the M&E plant should be supplied under a single turnkey contract.

However, a significant amount of civil works were involved including the embedment of the turbine spiral casing, the construction of the generator support as well as the equipment foundations in the switchyards at Gitaru and Kamburu. From past experience Mott MacDonald and Knight Piésold knew that few M&E contractors could complete these civil works themselves and would have to let subcontracts for their design and construction. Although the civil works were estimated to be only worth approximately 5% of the total project they would be firmly on the critical path for completion.
We therefore considered three options:

1. Complete turnkey contract including civil works design and construction.
2. Separate contracts for M&E and civil works (design and construction).
3. Single turnkey contract with civil design by MM/KPL

Most manufacturers of M&E plant have entered the turnkey contracting market as a means of keeping their factories busy. The staffs of these organisations often have a specialist background and can lack the cross-discipline experience that traditionally allowed consulting engineers to successfully integrate M&E and civil engineering works. Where new hydroelectric projects have been implemented with a single contractor, due to the value of his work it is usually a civil contractor that leads the turnkey organisation. Although contractors have successfully been able to put together M&E packages, their experience of civil engineering remains limited.

For Option 3 we anticipated that the tender documents would include reasonably detailed civil designs that would allow competitive pricing with only minor changes to suit the differences in equipment supplied by the various M&E contractors. These designs could have been quickly developed into construction drawings after contract award. This arrangement would allow the main contractor to concentrate upon the procurement of the M&E plant, as he would have done for many previous projects.

Although MM/KPL were confident of the reduction in engineering risk of them (and hence KenGen) retaining responsibility for the civil design, this work had not been included as part of the Engineer's scope. However, as long as contractors could be persuaded to tender for a complete package, then the financial risk to KenGen would probably be minimised with such an arrangement. We considered whether tenderers would increase their prices to take account of the additional risk but concluded that any such tendency would be minimised by the effect of competition. Furthermore we concluded that it would be difficult to convince KenGen that the main benefit of the Engineer with this arrangement was not an increase in the MM/KPL scope of work. Option 3 was therefore not recommended.

Although Option 2 would allow the E&M contractor to concentrate upon work that he was fully experienced in, due to the relatively small value of the civil works, it was concluded that there would not be possible to include sufficient monetary incentives or penalties to ensure prompt completion. This option was therefore not recommended. However, it subsequently became clear that by including the civil works as part of an overall turnkey contract this problem was not removed, but only transferred to the Contractor.

The recommended option was therefore Option 1 with the Contractor being responsible for the complete project. This arrangement transferred all of the design risks to the main contractor and thereby minimised the financial risks to KenGen.

Tailrace Tunnel

The original designers of the Gitaru scheme had recommended that the tailrace tunnel should be dewatered and inspected every ten years. However, this had not occurred since the project had been commissioned. Mott MacDonald and Knight Piésold strongly recommended that the tunnel should be inspected before the third machine was commissioned and the maximum flow increased by nearly 40%. KenGen accepted this recommendation and the tunnel inspection was carried out by the company that designed the original scheme over the Christmas-New Year period 1999-2000.

STAGE II

Programme

For most new hydro projects the overall programme is controlled by the construction of the civil works. The design and manufacture of the M&E equipment can therefore be completed without undue time pressures. However, for this project not only was the power station ready and waiting for the machine to be installed but the client was in a hurry. Completion in the shortest possible time was of prime importance.

In order to make tenderers keen to work quickly, adjudicated tender prices were reduced by 0.1% for every day that they were willing to guarantee completion in advance of the maximum period allowed. If tenderers were unable to guarantee to achieve the 700-day maximum specified period then their tender would be rejected. During the tendering period a number of the tenderers commented that this period was too optimistic and some were unwilling to tender if they had to guarantee such a short period. KenGen therefore reluctantly agreed to extend the target to 760 days. When tenders were received two companies guaranteed to complete the works in 670 days whilst two could only guarantee a 760-day period. The 90-day reduction in the guaranteed time for completion discounted the evaluated tender prices by 9% and this was significant factor that led to the award of the contract to Messrs Siemens and Voith of Germany, the same consortium that had supplied the original machines in the 1970s.

Civil Works Tender Programme

In order to take as much as possible of the civil work out of the critical path the Engineer proposed to depart from normal practice. The spiral casing is usually embedded and the concrete works forming the turbine floor completed before the generator support is constructed above. MM/KPL proposed that the generator floor should be constructed in advance upon columns that were extended from the bottom of the powerhouse. The spiral casing would then be constructed and embedded below the completed generator floor. Although this procedure would require temporary cranage to be installed below the generator floor for spiral casing erection the civil works would be completed earlier and this would allow turbine and generator erection to proceed in parallel in clean conditions. All tenderers based their tender programmes upon this sequence of operations.

S724/007/2000

Civil Works Implementation Programme

The Contractor's original programme followed a similar philosophy to that proposed by the Engineer. However, during the course of the project the programme reverted to a more standard procedure with the generator support being constructed after embedment of the turbine spiral casing. There were a number of reasons for this change. Firstly the Contractor had difficulty in awarding the contract for the price that he had included in his lump sum tender. Apparently there was a considerable difference between the two tender prices that he had received and he had based his tender upon the lower one. When he tried to award the contract based upon this figure the lower priced civil contractor raised his estimate such that it was similar to the higher tender price. The Contractor then re-bid the civil works and by the time the contract had been awarded it was too late to achieve the early start required. In parallel with these delays it is also clear that the Contractor was finding it difficult to manufacture some of the M&E plant in accordance with his tender programme and this was therefore going to be late on site. The delayed completion of the civil works was therefore not as critical as it appeared to be.

The effect of the delays in the civil works and the equipment supply meant that if the guaranteed completion date were to be achieved it would be necessary to compress the plant erection and commissioning.

M&E Erection Programme

The later than expected completion of the civil works as well as the late delivery of much of the M&E equipment meant that the Contractor had to work even harder than anticipated to achieve his guaranteed completion date. The main way that time savings were achieved was through parallel working. Gitaru has an underground powerhouse and space is limited but by careful planning it was, for instance, possible to build both the generator stator and rotor on the loading bay at the same time.

Assembly of Headcover and Regulating Equipment outside the Turbine

It is normal practice to erect the turbine internals in situ from the bottom up. The bottom ring and lower wearing ring are installed in the stayring and the lower guide vane stems entered in their bushes in the bottom ring. The runner is then centred in the lower wearing ring and the headcover lowered from above. The guide vane levers are then mounted onto the guide vane stems and regulating ring placed upon its bearings on the headcover. The levers and links connecting the guide vanes to the regulating ring are then installed.

In this instance the headcover was supported on a spare area of the generator floor. The guide vanes were then installed and temporarily supported from below. The guide vane levers were then installed thereby holding the guide vanes in place. The levers and links were then connected to the regulating ring, which had been placed on the headcover. When access to the spiral casing became available, the bottom ring, lower wearing ring and runner were installed before the complete headcover and guide vane assembly was lowered from above and the lower guide vane stems carefully entered into their bushes in the bottom ring.

We consider that the approach described above saved approximately 3 weeks in the erection of the machine.

Completion Date

Despite the efforts of all the members of the site team some equipment arrived on site too late and the machine was commissioned 30 days later than guaranteed. Only a few weeks before completion was achieved Mott MacDonald expected a much more significant delay but all parties involved, using a combination of flexible working and very hard work were able to complete the commissioning much more quickly than is usual. As some of the delays were for reasons beyond the Contractor's control it is likely that he will suffer no contractual penalty, although at the time this paper was prepared negotiations were still ongoing.

CONCLUSIONS

- The pro-active approach adopted by the implementation team on Gitaru (including the Client, Engineer and the Contractor) helped to achieve an ambitious implementation programme for the project.

- The Engineer's concerns regarding the experience of M&E contractors with civil works design and construction issues were justified as the changes required following design review were significantly more than those for the M&E supply. However, from an overall project perspective the problems with the civil works were probably no more significant than those associated with, for instance, manufacturing, shipping and customs clearance delays.

  We therefore consider that from KenGen's viewpoint a complete turnkey contract was the correct arrangement.

- The use and value of incentives to achieve an ambitious implementation schedule required to meet the energy generation deficit in Kenya can accomplish realistic results. It is unfortunate that the severe drought currently affecting Kenya has significantly limited the immediate benefits of this additional generation.

ACKNOWLEDGEMENTS

We would like to thank KenGen for their permission to publish this paper. We would also thank the staff of Mott MacDonald, Knight Piésold and Siemens and Voith for all their hard work in completing this project in a very short period.

© 2000 With Author

Beeston Hydroelectric Power Station development

S GOWANS
Hyder Industrial Limited, Runcorn, UK

SYNOPSIS

Beeston Hydro Scheme is one of the largest low head hydro schemes in the UK following construction completion in December 1999. The £3.5m project on the River Trent which has been undertaken by Hyder Industrial Ltd can generate 1.3 MW of electricity and reduce greenhouse gas emissions by some 5000 tonnes per annum from displaced fossil fuel generation.

The development is in a 'green belt', flood plain area and as a consequence has been constructed predominantly underground to prevent the obstruction of flood flow and minimise visual impact. This presented several design and construction issues which have been overcome by an innovative and creative approach.

The main construction area was surrounded by an enclosed coffer dam (Fig. 1) to minimise the likelihood of flooding during the construction phase. The scheme incorporates two bulb turbines manufactured by Esac Energie of France. Each turbine is capable under ideal conditions of developing 650kW of electricity. The scheme was procured using a multi contract strategy with Hyder Industrial managing all Contract interfaces and overall scheme programme.

1. INTRODUCTION

Prior to Hyder Industrial Limited becoming involved in the Beeston Hydro Scheme it had been developed by an alternative Company who found the Engineering and Regulatory restrictions with the scheme such that a commercially viable outcome was difficult to attain. Planning permission for the scheme had been granted to the previous developer but further consultation and development was required to ensure the scheme would provide the required commercial return and satisfy statutory and environmental standards. Following intensive engineering and commercial reviews Hyder Industrial Limited acquired the rights for the development of the Scheme at Beeston Weir in Nottingham under a NFFO Contract.

Hydro-electric power is generated by the use of the hydraulic head produced by the existing weir. This requires diversion of a proportion of the river flow around the weir and through the generation plant.

The turbine structure houses two bulb turbines with a maximum flow capacity of 74 m^3/sec, however at present it is only intended to operate to a maximum flow of 65 m^3/sec. Flow is diverted from upstream of the weir through a sheet pile wall inlet channel, incorporating a floating trash boom, bio acoustic fish fence and fish pass entrance.

The draft tubes discharge the flow from the turbine into the existing stilling pond (Fig. 3) adjacent to the side weir. This pond has been re-graded to accept the flow and redirect it back to the main river channel at the required velocity. The new profile of the pond will facilitate natural regeneration of gravel shoals providing a habitat for wildlife and preventing scour of the new river bank. Further measures, including the relocation of existing concrete blocks and sheet piling close to draft tubes have been provided to prevent erosion.

Two vertical bar screens are located on the intake to the structure to prevent debris from entering the turbine channels and damaging the turbines (Fig. 2). These screens are mechanically raked with a screen cleaning mechanism as and when required. The control system detects a loss of head across the screens and alarms to a remote 24 hr/day control centre to alert the site operators.

Flow control through the structure is provided by the use of the double regulated turbines. Stop log facilities are also provided to both the inlet and discharge to accommodate maintenance operations.

Water levels upstream of the weir will be maintained to ensure a flow depth of 75mm over the main weir as specified by the Environment Agency and as agreed by British Waterways to protect navigation entering the Beeston Canal.

2. DEVELOPMENT

2.1 Regulatory Issues
Many regulatory and planning issues needed to be overcome during the development stage of the Project. A selection of the issues are discussed briefly below :-

Noise emanating from the completed plant was raised as a Planning Condition. A background noise survey was completed prior to start of construction and will be repeated during operation. If required noise attenuation measures will be adopted.

In order to satisfy the requirements of the Environment Agency it was necessary to construct a new fish pass adjacent to the structure. The flow exiting the station draft tubes sets up an attraction to migrating fish hence a means of aiding migration upstream was required at that point. In order to prevent the fish entering the hydro station draft tubes appropriate barriers upstream and downstream have been installed, these are discussed in detail below.

A temporary discharge consent was required during the construction phase of the project to ensure ground water pumped out of the excavation was to a satisfactory standard prior to return to the water course. The required water quality was achieved via the use of a temporary stilling pond constructed for this purpose. The stilling pond was effective and maintained the required quality to the satisfaction of the Environment Agency.

2.2 Land Issues and Access

Access to site under the previous development agreement and Planning permission was through Holgate Village, a small picturesque village with narrow roadways with final access to the River Trent via a hollow road and bordering several Sites of Special Scientific Interest (SSSI). This access and egress route rightly enforced restriction on vehicular movement to minimise disruption to residents. It was therefore necessary to locate alternative routes to the site that would enable free flow of construction traffic. An alternative route was located and secured across fields from the main A453 carriageway.

3. PROCUREMENT AND CONSTRUCTION

3.1 Design Development

Following review of several turbine types, configurations and manufacturers which included maximising output and minimising cost the option chosen was two ESAC bulb turbines as described in section 3.3 below. Several iterations of capital cost versus income were performed to determine the most economic solution. Analysis was performed on sixty years of historic river flows obtained from Walinford and the flow available for 65% of the time was utilised as the base flow figure from which turbine size and configurations were determined. The conclusion was a twin turbine configuration to allow two turbines to be utilised for the 65% time frame and a single turbine during the lower 35% of the time, this ensured best use of turbine efficiency and output.

The civil design was awarded to Hyder Consulting Limited who assisted in providing civil cost analysis for the various turbine configurations under consideration and then onto full detail design of the civil works required for the chosen option.

Intake configuration was modeled to determine the most appropriate dimensions to suit the turbines under consideration. Velocity profiles were plotted to aid this process. These velocity profiles were also utilised in determining the optimum location for the upstream fish screening system. The sections have been installed on a velocity profile of 0.6 m/sec to enable the fish to safely swim away from the units. Higher velocities could be such that the fish are unable to swim against the resultant flow.

3.2 Contract Strategy

It was concluded during the design development and cost analysis that the most secure route for Hyder Industrial Limited to procure and construct the facility was to utilise a multi-contract strategy as opposed to a turnkey package. This provided flexibility in terms of Standard Forms of Contract used for all elements of the plant and the ability to optimise cost reductions and savings during the late design stages and throughout construction. The scheme was split into the following major Contract elements :-

1. Civil design
2. Civil construction
3. Turbine package (including design, construct, install, commissioning)
4. Electrical works (including instrumentation, MCC, site wiring, commissioning)
5. Fish protection equipment (upstream and downstream)
6. Other minor Contracts

3.3 Plant and Equipment

3.3.1 Civils
The civil works were designed specifically to accommodate the two ESAC turbines and to comply with the stringent Planning requirements, ESAC provided specific civil details for elements of the works that needed to be accommodated within the overall design. This included intake velocity, draft tube geometry and other detail related to the flow velocity through the system and turbine dimensions.

The site is located in a flood plane and hence minimal structures were allowed above flood plane / ground level. This issue in itself provided difficulties in design and installation with regard to confined space entry and water ingress. The control room and distribution chambers were designed and constructed to be water retaining due to the location below flood level. All duct entries into the control
room were carefully planned prior to construction and utilised individual sealing mechanisms for each cable entry into the structure.

3.3.2 Turbines
Two ESAC bulb turbines (Fig. 2) are now in operation a summary of their specification is detailed below :-

Net head (m) (designed)	2.04
Runner speed (r/min)	110
Maximum flow (m^3/s)	37
Runner diameter (mm)	2780
Number runner blades	3
Number of guide vanes	16

The theoretical efficiency of the turbine at design Net head and maximum flow is 88.90% with a maximum efficiency of 92.10% at 22.2 m^3/s at the design net head.

The generator details are as below :-

Manufacturer	Leroy Somer
Nominal output	702 kW

A step-up transformer is used to convert the voltage from 440 V output at the generator to 11 kV for transmission to the local grid. The step-up transformer is located at the hydro plant and a 1 km cable transfers the power to the regional electricity network.

3.3.3 MCC and control system

The MCC and control system was designed in conjunction with the turbine manufacturers requirements and the necessary controls required to maintain river levels and flows. Following many design reviews the final package was implemented and a complete system functional test was undertaken at the manufacturers works prior to delivery to site. The MCC and control system package included all control and power requirements for both turbines and all ancillary plant and equipment. An Allen Bradley SLC 5/04 Programmable Logic Controller (PLC) has been utilised together with a Panel View 550 operator interface unit for display of all parameters and set points which can be altered following password input.

3.3.4 Fish protection equipment

As part of the consultations with the Environment Agency it was concluded that a fish screen would be required both up and downstream of the hydro station. Due to the scheme being low head it was imperative to find a system that would minimise the head loss through the system. The conclusion was the installation of a Bio Acoustic Fish Fence (BAFF) up-stream of the intake and a Graduated Fish Field Barrier (GFFB) at the outlet from the draft tubes supplied by Fish Guidance System Ltd.

The BAFF uses compressed air to produce a continuous curtain of bubbles around the intake to the station. The curtain rises from the river bed and encapsulates acoustic sound waves set-up by driver units located within the river bed sections. This resultant wall of sound to which the fish are sensitive (10 - 1000 HZ range) guides the fish away from the high velocity intake currents and towards the fish pass which has been installed adjacent to the plant.

The GFFB consists of a series of electrodes built into the circumference of the draft tube walls, the electrodes are flush with the wall and hence provide no restriction to flow through the plant. A programmable output waveform pulsator introduces current to the electrodes introducing an electric field between them. When fish enter this field they become part of the electric circuit with some of the current flowing through their body. As the fish advance into the graduated field they feel an increasingly unpleasant sensation, when the sensation becomes too great the fish are swept out of the draft tube to safety.

3.3.5 Other plant

A floating trash boom has been installed across the intake channel to divert gross debris away from the intake channels, this boom consists of several steel tubular sections with 500mm skirts. Navigation buoys have also been installed across the river to warn boating traffic of the intake to the structure.

A forced ventilation system has been installed in the control room to reduce heat build-up from the transformer, air blowers and panel and also to provide adequate ventilation for personnel access and minimise the confined space categorisation.

3.4 Site Co-ordination

The civil contractor was appointed Principal Contractor under the CDM Regulations 1994 and coordinated day to day site activities and safety issues under the management and supervision of Hyder Industrial Ltd.

Hyder, as Construction Managers, maintained full responsibility for the inter contract interface management, programme and budgetary controls. Fortnightly site construction and design coordination meetings were instigated throughout this 'fast track' project and at each meeting procurement status reports were issued and approved to ensure minimal abortive or non-productive working. Through these regular reviews of progress and proactive site management the civil structure was completed to the initial project programme in order for turbine and ancillary plant to be installed.

The installation of the turbines and ancillary plant progressed well and was completed to enable commissioning and testing of the station early in December 1999.

4.0 PERFORMANCE AND MONITORING

4.1 Performance Tests
Detailed performance tests are to be carried out on the plant following completion of this paper. Limited data has been collected to date which indicates that output from the plant is as expected and income has exceeded expectation during the summer months.

4.2 Output Review
A review of output against the predicted average annual energy production will be undertaken when the detailed performance tests have been completed. Early indications are that the output will be as predicted.

4.3 Telemetry
The plant has been designed to be unmanned for the majority of the time with occasional operational visits for trash cleaning of the inlet screens and physical inspection and monitoring of the plant and equipment. The site is remotely monitored on a 24 hr basis by Hyder's Control Center which alerts the required personnel should any out of specification or trip alarm occur. The system logs all turbine safety parameters such as bearing temperatures, oil temperatures and pressures, generator winding temperatures, plant output and drive failures. The outstation utilised is a Servelec S500 unit which receives alarms and parameter data directly from the control system PLC, these alarm conditions together with regular data updates are passed to the control centre via a standard phone line.

4.4 Reliability
The plant to date has suffered minor failures which have not significantly affected the long term energy output of the scheme. The main issue to date relating to maintaining plant output, has been the requirement to regularly monitor and calibrate instruments that provide turbine blade position feedback to the control system, the position feedback to the control system was indicating turbine opening 20% higher than actual, therefore when the feedback to the control system indicated maximum opening (100%) the actual opening was approximately 80% hence reducing system output.

4.5 Trash Cleaning
The plant performance can be significantly affected by the build up of debris on the inlet screens to the station. Due to the system being low head (approximately 2.5m average net head) any loss of flow or head across the system can reduce power output substantially. Due

to the limitation of above ground structures and equipment it has been found that proprietary trash cleaning mechanisms have been unsuitable. The quantity of material to be encountered was an unknown and it was therefore decided to utilise a mobile cleaning mechanism as required during the initial operation of the scheme with a view to evaluating the need for a fixed system in the future. A design for a fixed system has been developed in preparation but early indications are that the mobile method is sufficient which saves on the substantial capital outlay.

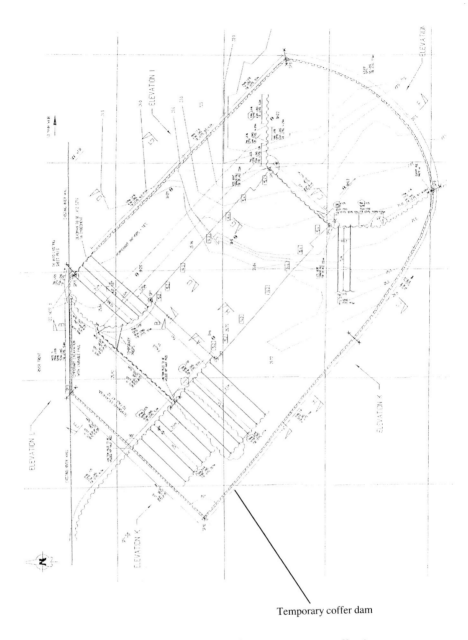

Temporary coffer dam

Fig 1. Plan view of construction area showing temporary coffer dam

S724/008/2000 © IMechE 2000

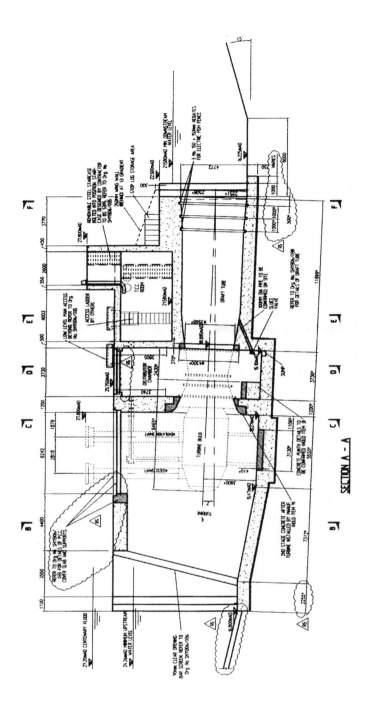

Fig 2. Section through turbine structure

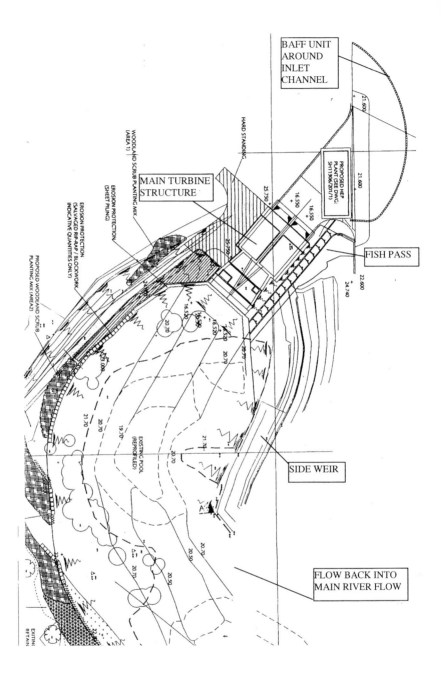

Fig 3. Plan view of structure in relation to side weir and down stream pond

S724/008/2000 © IMechE 2000

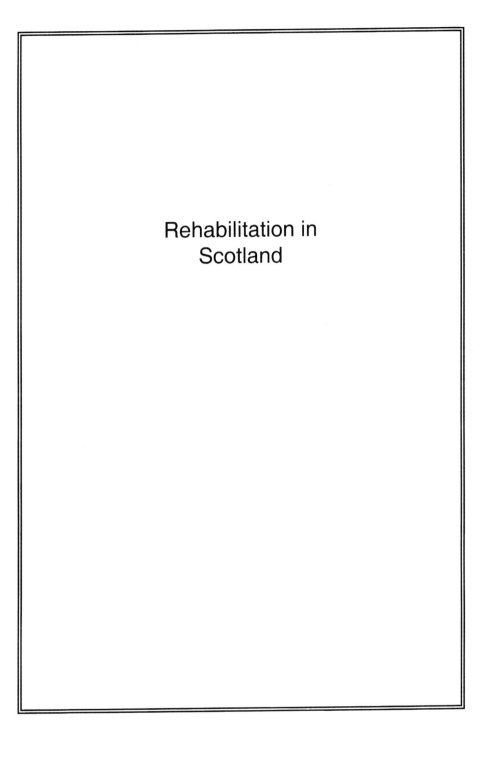

Rehabilitation in
Scotland

S724/009/2000

A strategy for the refurbishment of a 1080 MW hydro generation portfolio

J I SMITH
Scottish and Southern Energy plc, UK

SYNOPSIS

Scottish and Southern Energy own and operate 1080MW of conventional hydro and 300MW of pump storage plant, supplying over 40% of the UK's renewable energy. The vast majority of the conventional hydro was constructed in the 1950's or earlier. This paper discusses the strategy which has been formulated to refurbish the portfolio and the progress made since work commenced in 1997.

1. INTRODUCTION

Scottish and Southern Energy's (SSE) 1080MW of conventional hydro generation with an average annual output of 3200GWhrs is located across the north of Scotland. The portfolio is characterised by the large number of small/medium sized stations with extensive civil works to divert and collect water. Figure 1 provides an illustration of the number and location of sites and the associated catchment areas.

Control of the plant has been progressively centralised to the point where all the stations, other than a few small sites, are now under full remote control from a single control room located at Clunie Power Station.

In 1997 SSE set out to extend the life of its ageing hydro generation plant and increase the annual production through improved efficiency and reliability. To achieve this it adopted the following strategy:-

* Fully refurbish 30 stations within a 15 year period.
* Each refurbishment should put the station to a state which gives the expectation of 30 years reliable service without further requirement for significant capital expenditure
* Station refurbishment contracts to be awarded on a turnkey basis.

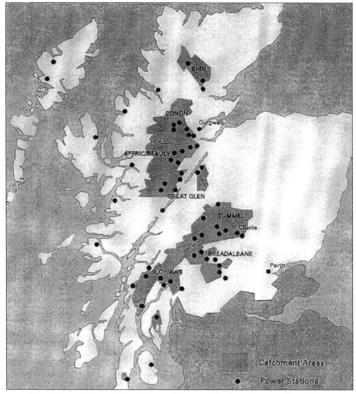

Fig 1 Scottish and Southern Energy Hydro Generation Sites

The overall objective of all the above was, of course, to maximise both the profits of the hydro generation business within SSE, and the amount of sustainable/renewable energy produced.

Although the strategy is a simple one, it requires substantial capital investment. It was only the operational and refurbishment experience of previous years that provided the encouragement and justification to pursue such a bold strategy.

2.   PLANT HISTORY

SSE's hydro generation plant dates back to 1930, when the Grampian Electricity Company constructed two stations at Rannoch and Tummel Bridge with a combined output of 76MW. When the UK Electricity Supply Industry was nationalised in 1948 ownership of these stations transferred to the North of Scotland Hydro Electric Board (NSHEB), the predecessor of Scottish Hydro Electric and latterly SSE.

In 1943 NSHEB had embarked on a major hydro construction programme. The first new main station to be constructed was opened at Sloy in 1950 . By 1963, about half the estimated hydro-electric potential of the Highlands had been developed. Fifty four main power stations had been completed with a total generating capacity of over 1000MW. Fifty six main dams had been built, some 300km of rock tunnel excavated and a similar length of aqueducts and pipelines constructed. This growth in installed capacity is illustrated in fig. 2.

Despite the fact the technology employed was new and the construction programme short, the plant performed very well on the whole, which is a great tribute to all those employed in their original development and construction.

By the late 1970's a targeted programme of plant overhauls and upgrades was underway to arrest deterioration and address known problems. Despite this programme of work, significant plant failures were beginning to surface and were becoming more frequent. It was clear that what had started as a planned incremental approach to refurbishment, was by necessity becoming more reactive.

While it would have been possible to increase the refurbishment work undertaken using this incremental approach. Experience had shown the strategy to have a number of serious shortcomings, which undermined the perceived savings of delaying investment across the whole station. The main disadvantages are summarised below:-

•   The financial benefits of discrete investments were not being fully realised as the reliability and performance were constrained by other un-refurbished plant within a station.
•   Discrete refurbishment works led to multiple outages reducing the availability of the plant on an ongoing basis.
•   The ability to optimise the plant performance when integrating new plant with old is not always possible.
•   The engineering resource required is greater and therefore more costly.
•   It is a time consuming process which as will be discussed later was becoming a key issue in SSE's case.

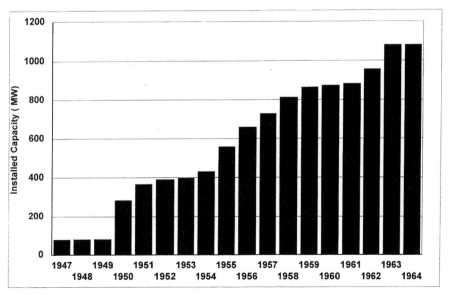

Fig 2 Growth in Hydro Generation Capacity

By the early 1990's the above shortcomings were becoming more prominent and a turning point was reached following a generator fire on one machine at Sloy. The turbine runners had only just been replaced at Sloy following a previous turbine failure. The generators were at the time considered to be in a reasonable condition and no work was planned at that time. This subsequent failure served to illustrate the cost of not refurbishing the complete plant together, and that although individual plant items could continue to operate for a period of time without failing, the plant had in general terms reached the end of its life. At this point the decision to completely refurbish Sloy was made.

Although this was a positive change in approach, it brought with it its own new set of problems. To achieve this complete station refurbishment discrete contracts were awarded which were divided in much the same manner as they would have been had an incremental approach been adopted. This resulted in over 30 different contractors being employed in parallel.

Apart from the complexity of co-ordinating the activities of so many different groups and integrating the different plant and equipment, it was impossible to have adequate commercial protection for under performance across contracts. Delays in one area became compounded with little redress to recover the resulting financial loss.

It was also apparent that the defined scope of refurbishment needed to consider the impact on the complete project and not just the cost of any particular element. In the past small discrete jobs would require plant to be stripped and examined before establishing the full scope of work. The consequent delays which can arise as a result, can in many instances have a disproportionate affect on the cost of the total project. Added to the fact that for the size of the plant involved replacement rather than repair is often more cost effective and that the plant is now 50 years old, a much more aggressive approach to the refurbishment scope needed to be employed.

In 1996 a complete review of the long term refurbishment requirements and how these would be achieved was undertaken. As well as taking account of the refurbishment experience to date, another key point highlighted the need for change. An assimilation of the complete portfolio's age adjusted to account for past reinvestment highlighted a disturbing trend. See fig 3.

Despite a steady capital investment in refurbishment since the late 1970's, the average plant age was still increasing. Based on the known condition, it was clear the plant was reaching the end of its economic life and the increasing plant age should be reversed if the long term future of the stations was to be protected.

The strategy employed today was formulated from this review. The impact this strategy will have on the future average age of the plant is also illustrated in fig 3.

S724/009/2000

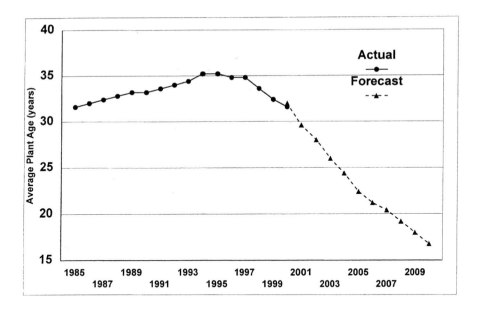

Fig 3 Historical and Forecast Average Plant Age

3.   PROJECT ECONOMICS

With the strategy now in place it was still necessary to demonstrate on a case by case basis the economic benefits of refurbishment.

SSE's hydro plant is characterised by its small/medium size and low/medium head combined with a relatively low load factor. The average load factor being only 35%. None of the above are conducive to achieving a good return on any investment in the plant. Indeed it is unlikely in the current UK market that the stations could be constructed on an economic basis today.

Although the cost of refurbishment is substantially less, the economics remain fragile. The investment returns of each refurbishment project are rigorously scrutinised and must achieve the necessary hurdle rate to obtain funding.

While the station refurbishment projects undertaken to date have all achieved the required investment return criteria, other stations which are known to be less profitable may prove difficult to justify. This may be addressed by existing renewable generation attracting a premium for its output in the future. However it appears that hydro plant over 10MW capacity (defined as large hydro) may be excluded from any legislation, which could threaten the long term viability of these stations once refurbishment becomes essential, and could also seriously affect the Government's renewables targets.

4. PROGRESS TO DATE

Since 1997 five station refurbishment contracts have been let: Glenmoriston (40MW); Rannoch (42MW); Errochty (75MW); Pitlochry (15MW); Clachan (40MW), at a total cost of over £35M.

The first two contracts at Glenmoriston and Rannoch have been completed with the next three all due for completion in the 2000/01 financial year. The next two contracts for Inverawe (25MW) and Gaur (6MW) are under development and should be awarded by the end of 2000.

5. THE GLENMORISTON PROJECT

Details of the first project at Glenmoriston provides an insight into the work undertaken during a station refurbishment.

Glenmoriston is a 40MW station located on the River Moriston 4 miles north of Loch Ness. The station comprises two 20MW vertical Frances machines operating at a gross head of 97m. The machines are located in an underground cavern which is set directly below the head pond formed by the Dundreggan Dam. See fig 5. Water is discharged by a 4 mile long tailrace tunnel into Loch Ness. The main storage for the station is Loch Loyne and Loch Cluanie which discharge through Ceannacroc Power Station into Loch Dundreggan

The average annual output of the station since it was first commissioned in 1957 is 110GWhrs. After 40 years of operation the condition of the plant was beginning to deteriorate and much of the equipment was obsolete. In the year prior to refurbishment two separate circuit breaker faults had led to long outages as no spares were available.

5.1 Programme

Tenders for the complete station were invited at the beginning of 1997. The contract was awarded to Alstom Hydro Ltd of Rugby in July 1997. Work at the Dundreggan Dam commenced a month later with the first machine being taken out of service in February 1998, followed by the second machine in April 1998. The first machine was completed in January 1999, followed by the second machine in February 1999.

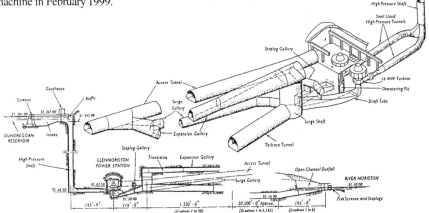

Fig 4 Glenmoriston Power Station and Associated Tunnels

5.2   Scope of Work
The scope of the work was extensive as illustrated by the list below:-

* Refurbished turbine with new runner, guide vanes and cheek plates.
* Refurbished generator including re-cored and rewound stator, re-insulates rotor, new brushless exciter and new water cooling system.
* New plc controls, including digital governor, AVR and electrical protection.
* New 11kV circuit breakers
* New 400v distribution system
* New lubricating oil system incorporating a high pressure jacking oil system.
* New high pressure governor oil system and servo system
* New DC systems
* New pipework and cabling
* New drainage system including oil separation plant
* New station ventilation system
* New intake screens and screen cleaning plant
* Refurbished intake gates and draft tube gates
* Refurbished main dam flood gates and new automatic controls
* Refurbished Borland fish pass gates and new controls
* Refurbished dam trash and scour gates
* Refurbished compensation set and new controls

It can be seen that apart from the large scale high value plant such as turbine casings, rotors and large gates, the majority of the equipment was replaced. The merits of replacement versus refurbishment are considered on a station by station, item by item basis and will therefore vary between projects.

5.3   Plant Efficiency
As well as restoring the station to an "as new" condition, the opportunity was obviously also taken to maximise the efficiency of the plant.

A study confirmed the existing rating of the station was well matched to the hydraulic system and catchment and any increase would only serve to reduce the annual income of the station. Consideration was also given to lining the long tailrace tunnel to reduce friction losses, however this proved to be uneconomic.

The original turbines at Glenmoriston had not been optimally matched to the site conditions. The turbine best efficiency point was at a lower net head to enable the same turbine to be installed at both Glenmoriston and nearby Ceannacroc which has a lower head, making the original turbine selection a compromise.

The new turbine has been matched specifically for Glenmoriston which combined with advances in design and refurbishment of other worn components, has resulted in a 10% increase in turbine generator efficiency. The overall performance has been verified by site measurement where an efficiency of 91.5% was recorded.

5.4   Contract Conditions
The contract was let under the New Engineering Contract conditions with guarantees for both programme duration and plant performance. This was the first time either party had used

these conditions on a project of this size or complexity. It is certainly the view of SSE that the use of these conditions was beneficial, helping to ensure problems were addressed quickly as they occurred while maintaining a strong focus on successfully completing the project.

5.5 Early Operating Experience

The plant is now well through its 24 month warranty period. Although the plant has experienced a number of minor problems, which are to be expected following such extensive works, the overall reliability of the station has been very good.

The improved efficiency of the machines has been reflected in the annual energy produced which saw a 10% increase in the first year equating to an additional 11GWhrs.

6. CONCLUSION

After the first four years of implementing an intensive and comprehensive refurbishment strategy, many of the perceived benefits are now being realised. With 230MW of capacity either refurbished or nearing completion preliminary work has already commenced to refurbish a further 200MW over the next few years. The performance of the portfolio as a whole is improving and most importantly the life expectancy of the plant is now no longer in decline.

SSE will continue to monitor the performance of each project and learn from any lessons learned, however based on the experience gained we believe the basic strategy to be sound and only the detail of how it is achieved needs refined.

Some areas where improvements may be possible and which are being considered are the application of partnering with contractors and nominating preferred suppliers for key equipment to avoid proliferation.

One area of concern is that current proposals classify large hydro (over 10MW) alongside other non- renewable forms of generation, affecting the long term profitability of the stations within the UK market. Without the necessary recognition of the role existing hydro generation plant plays in contributing to the reduction of greenhouse gases, some stations may prove uneconomic to refurbish and raise the possibility that stations may be decommissioned in the future.

© 2000 With Author

S724/009/2000

The refurbishment of Rannoch Power Station including replacement of the spiral casing

N KIDD and **N J WHYTE**
GE Energy (UK) Limited, Doncaster, UK

1. INTRODUCTION

Rannoch Power Station is nominally a 45 MW output, three unit hydro generating station operating on 144m head originally commissioned in the early 1930's. The station is located in Scotland in the Tummel valley which runs from the famous Rannoch moor above Glencoe on the West coast to Pitlochry in Strath Tay in the East. The headwater for Rannoch Power Station is taken from Loch Ericht, a high and remote reservoir in the central Grampians. Rannoch is used as base load for the Tummel group of stations, black start facilities being provided from elsewhere.

The station owners, and therefore the client, are Scottish and Southern Energy, formerly Scottish Hydro Electric. This company has embarked on an ambitious 15 year refurbishment plan to ensure that hydro power remains at the centre of Scottish power production whilst offering maximum returns to their shareholders.

2. TENDER PREPARATION

GE Energy, formerly Kvaerner Boving, received an enquiry for competitive tender for the refurbishment of Rannoch in 1997 and spent several years developing the project in conjunction with the client to achieve their best return. As such many options were evaluated. The original specification given by Scottish and Southern Energy asked for a replacement runner and guide vanes only keeping the spiral and stay bolts. The stay bolts consisted of pairs and single bolts through the spiral that held the spiral casing together. Furthermore the original specification asked to refurbish the spiral relief valve. Operating experience from as early as the original commissioning had shown problems with penstock resonance which resulted in further work by the client and ultimately a load restriction and maximum penstock pressure being imposed. These restrictions were to be maintained as refurbishment design conditions along with a vibration study to ensure that the new design did not increase the vibration.

During preparation of the bid the hydraulic transient study showed that in the case of relief valve failure very large over pressure in the penstock would occur. The final sections of penstock between valve house and powerhouse are exposed and therefore failure could have catastrophic results. To reduce the penstock pressure rise a long guide vane closing time would be required, consideration of which led to the proposal of removing the relief valve with the consequence of reducing the amount of maintenance required on the machine and increasing its safety. In removing the relief valve the spiral could also be replaced allowing a further improvement in the machine efficiency.

Losses were estimated for the plant as it was designed originally and compared with modern components fitted into the available envelope, see fig1. Replacement of the spiral which required changing stayvanes and removal of the relief valve increased efficiency about 1.2%, a significant improvement. Things that could not seriously be altered, draft tube length, spiral inlet position, spiral inlet diameter, and machine centre line were maintained and CFD was used to optimise runner inlet height for the spiral diameter and new guide vane profile and stay ring. Runner outlet diameter was reduced and the suction cone angle was changed to increase the efficiency of the draft tube. With the spiral out the draft tube could be more easily modified to improve efficiency further with a new draft tube profile constructed inside the original shape.

Several different design options were presented to the client. Problems were highlighted in keeping the existing spiral since this reduced the potential efficiency of the new machine, introduced problems of over pressure and presented a risk of an extended outage if significant work was required on the existing stay bolts following NDT. It was not possible to perform NDT prior to a major outage since some concrete had to be removed to allow access to all the bolts and assessment of the condition of the studs was difficult as defects would probably be at the root of the first few load bearing threads. These bolts had also been in place for approximately 70 years and subject to vibrations caused by vortex shedding in the flow together with stress reversals from pressurising the spiral casing each time the turbine started. This coupled with lack of bolt pre-load suggested that they would also have a reduced fatigue life. Replacement would also be difficult if found to be defective as the threads at the upper and lower ring were made as one continuous thread. Also, they reduced the efficiency still further since a non-optimal runner inlet height had to be selected if the original guide vane height was used. So with new spiral GV height increased 20mm or so, combined with10 stay vanes and 20 guide vanes allowed for an optimal hydraulic design solution.

Further more problems were expected in welding in fairings which were proposed to fit round the stay bolts to give a better hydraulic profile.

A new spiral allowed maximum efficiency given the conditions which remain fixed, including replacement of the draft tubes, removal of the relief valve to increase safety and optimisation of the remaining turbine components. These included a new external regulating ring with smaller servos, and a runner pumping plate design new to the UK to eliminate the need for external cooling water pump sets.

3. VIBRATION STUDY AND DESIGN CONSIDERATIONS

Scottish and Southern Energy accepted the arguments for improved performance and safety and opted for a new spiral and draft tube design and awarded the Contract to the Kvaerner Boving – Elin Consortium in December 1997.

An initial vibration study of the penstocks was performed in the winter of 1998 which showed that the problem was not wake shedding from the penstock protection valves as originally thought but was due to the draft tube pressure pulsation at overflow.

Once confirmation of the scope of work had been agreed the main areas of mechanical consideration centred around the effective re-planting of the turbine into the given envelope of the existing plant. This had to incorporate both the new hydraulic and mechanical design within the civil constraints of the building.

One of the major challenges was associated with the spiral thrust collar & the thrust loads on the civil works. The original spiral casing had been of a cast construction and was considerably thicker in section than a new casing designed to BS5500. The potential consequences of this were that a thinner casing whilst able to withstand the pressure loading contained within might add some extra concrete forces due to potential expansion of the casing. The new arrangement consisted of a dismantling joint downstream of the inlet valve so that all spiral casing forces had to be contained within the immediate vicinity and there would be little concrete cover to support it. The solution chosen for the final design had a thrust collar on the spiral with the thrust taken by a heavily reinforced concrete beam transferring the load to mass concrete. In parallel with this the casing was covered in the appropriate areas with a thin flexible membrane to minimise load transfer to the concrete structure.

Other mechanical consideration included:

- The final design has small angular offset of spiral and skirt to inlet and draft tube due to a reduced spiral diameter.

- Consideration of loss of centre line datum. Normally the draft tube would form the reference and the unit would be built up from that position. In this case with no draft tube, spiral etc the original centre had to be found and used as reference. As a guide, the generator reused the original thrust brackets which only had a very limited degree of adjustment and therefore all alignments issues had to be conducted with great accuracy.

4. SPIRAL MANUFACTURE

The first spiral was manufactured in eight months including works build of the turbine. Pressure testing was done in works to minimise the site programme. Painting was carried out in works. Delivery to site was with special transport requiring police escort due to it being an oversize load.

5.   SITE INSTALLATION

Once it became clear during the tender stage that a replacement spiral was viable alternatives were established for installation. The programme did not allow for the spirals to be installed in sections and either welded or bolted and then pressure tested. This presented a further problem in that the spiral would have to be embedded empty rather than the current practice of embedding whilst pressurised.

The initial favoured installation method was to break out the walls at turbine level and drop the spiral onto cantilevered beams over the tailrace. After contract award a full civil survey of the station was performed to establish the existing condition. Unfortunately this showed that the bridge on the south side of the station over the tailrace deck had been significantly damaged by frost and water and could now only maintain a much reduced load capacity. An extremely large crane would have been required to lift the furthest spiral into position and would have been prohibitively expensive. The original installation plans were therefore abandoned.

The next alternative was a Bailey type bridge, but the span and loads involved again proved too much for a crane to sensibly install. Whilst driving back from a project meeting in Scotland a couple of project members were stuck in a traffic jam under some motorway bridges under repair. A bit of lateral thinking after observing the support systems used resulted in the idea of a temporary lightweight bridge structure across the tailrace being supported off the tailrace deck. The concept of dewatering the tailrace to allow access to the draft tubes for replacement had already been considered, and now with the ability to install a bridge at the same time, the concept was resurrected.

The programme allowed a short outage at the end of 1998 to install new tailrace stoplogs following which two of the three machines had to remain in service until summer 1999 when a total station outage of eight weeks was programmed. The programme involved replacing the first turbine and then the following two in parallel with about one month stagger. Obviously the temporary bridge had to be in place for all three units so that the spiral could be installed for one unit whilst the other two were operational. Thus any bridge had to be installed in August 1998 and be removed prior to the end of September 1999 when the complete station outage finished. A Mass 50 temporary bridge designed by Mabey Support Systems and as used on many motorway bridge repairs was installed and successfully used. The design was based on placing the spiral onto the end of the bridge supported by the main banking and then dragging across the bridge on special skates designed and manufactured to suit the bridge beams. Once opposite the appropriate unit position, the spiral was jacked up and the skates turned through ninety degrees and then the spiral pulled into the power station and over the pit. The spiral was then lifted through the aperture created by the dismantled generator and lowered into position.

As already described the positioning of the spiral became the complete machine centreline since it was installed prior to rebuilding the draft tube due to programme constraints.

The installation of the spiral into each pit took a day per machine and was completed without problems. The method of concrete embedment was successfully achieved and no cracking has been observed. The final machine at Rannoch went into service one day ahead of programme and all performance and commercial guarantees were met.

GE Energy and their partner have proved that with a high calibre development, design and management team, the refurbishment of old hydro power stations can be made commercially viable and completed on time and to budget. The application of modern technology to these old stations is achievable and can produce impressive improvements in performance and efficiency whilst reducing maintenance.

Comparison between new and refurbished components	
Turbine Component	Improvement In Performance (%)
Spiral casing	0.1
Relief valve	0.15
Stay vanes	0.95
Distributor	0.5
Runner	1.1
Draft Tube	0.8
Other	1.43
Total	5.03

Figure 1

Figure 2 General view of generator hall with unit 2 stator dismantled

Figure 3 Original turbine, spiral manhole and flanges just visible above concrete

Figure 4 Commencement of coring around turbine

Figure 5 New spiral in place with rebar for thrust beam

Figure 6 Spiral embedded awaiting MIV make up piece and GV apparatus

Figure 7 Unit 2 complete

© 2000 GE Energy (UK) Limited

Authors' Index